SpringerBriefs in Environmental Science

For further volumes:
http://www.springer.com/series/8868

Xiangzheng Deng · Yi Wang
Feng Wu · Tao Zhang
Zhihui Li

Integrated River Basin Management

Practice Guideline for the IO Table Compilation and CGE Modeling

Springer

Xiangzheng Deng
Tao Zhang
Institute of Geographic Sciences and
 Natural Resources Research
Chinese Academy of Sciences
Beijing
China

Yi Wang
Institute of Policy and Management
Chinese Academy of Sciences
Beijing
China

Feng Wu
State Key Laboratory of Water Environment
 Simulation
School of Environment
Beijing Normal University
Beijing
China

Zhihui Li
Institute of Geographic Sciences and
 Natural Resources Research
Chinese Academy of Sciences
Beijing
China

and

University of Chinese Academy of Sciences
Beijing
China

ISSN 2191-5547   ISSN 2191-5555   (electronic)
ISBN 978-3-662-43465-9   ISBN 978-3-662-43466-6   (eBook)
DOI 10.1007/978-3-662-43466-6
Springer Heidelberg New York Dordrecht London

Library of Congress Control Number: 2014939384

Printed on acid-free paper

Springer is part of Springer Science+Business Media (www.springer.com)

Preface

Since the problem of natural resource depletion and the economic development is at the crossroad of sustainability, how to efficiently and reasonably exploit and use the limited resources to sustain the balance has become an urgent governmental and scientific topic related to the entire human society and natural ecosystems. In the former researches, hydrological models are mainly used to explore reasonable water management, considering water as a natural element rather than a social-economic one. However, the social function of water should be taken into consideration seriously when policy makers look forward to construct an integrated management system. The purpose of this work is to propose a conceptual framework of regional input–output table compilation at regional level and introduce how to incorporate the Computable General Equilibrium (CGE) model into integrated water management research to explore a more sensible and optimal method to implement sustainable development for the river basin.

Establishing multiscale optimal water resources allocation modes which takes the enhancement of utilization efficiency as core is a research hotpot for the international water resources management of river basins. From the perspective of data integration which serves the management of water resources, this work aims to introduce how to compile the first set of regional level integrated IO tables involving resources and environment accounts with integrated datasets which contains the spatio-temporal data of water and land resources, ecology, and social economy in the river basin, and how to construct an integrated CGE model with resources and environment accounts embedded, which can be used to quantitatively depict the key process parameters of the water-ecology-social economy coupling system. Thus, this work can provide decision support for integrated river basin management, and scientific support for the sustainable development of social economy, eco-environment, and water resources.

Theories and methods to address this issue are supposed to distinguish and analyze the interrelationship and mechanisms within these complex systems. Among which the input–output analysis is an economic method for analyzing the interdependence of an economy's various productive sectors by regarding the product of a particular industry both as a commodity demanded for final consumption and as a factor input in the production processes of itself and other sectors. Input–output tables, as the core of input–output analysis, can be constructed for the whole or segmented economies in planning the production levels to meet the consumption demand and in modeling the impacts of economic activities such as component changings.

An overview analysis of existing challenges and opportunities in certain resource-restricted areas associated with the integrated management finds out that there is considerable potential for regional green development as well as various severe issues need to be addressed. The conceptual framework is mainly focused on two natural resources including the water and land resources. How to construct an integrated CGE model is explained and the implementation of an integrated CGE model with TABLO language is displayed. The integrated CGE model can be used as a tool in the analysis for enhancing the resource security in particular regions such as inland river basin and resource-restricted developing areas. This work can also provide insight into bridging the gap between national and small regional input–output analysis.

Several chapters and sections include concrete examples. More assistance from the relevant literatures can also be found in the references at the end of the document.

The authors claim full responsibility for any errors appearing in this work.

April 2014   Xiangzheng Deng
                                                                          Yi Wang
                                                                          Feng Wu
                                                                          Tao Zhang
                                                                          Zhihui Li

Acknowledgments

This work was financially supported by the major research plan of the National Natural Science Foundation of China (Grant No. 91325302), the National Natural Science Funds of China for Distinguished Young Scholar (Grant No. 71225005), and National Key Programme for Developing Basic Science in China (Grant No. 2010CB950900). We would also like to thank our project team who should be the most knowledgeable for their unwavering support throughout the production of these results. Without their contribution, this work would never be published. Finally, we are grateful to the Springer staffs for their lasting enthusiasm to help us produce this work.

We are grateful to all the authors of the numerous books and research publications mentioned in the list of references at the end of this work. These valuable literatures formed the foundation of this work. We express our gratitude to those researchers and organizations for their contributions that reinforced our knowledge.

Steering Committee

Contents

Chapter 1
Introduction and Overview

1.1 Background of Integrated Water Management

With an increasing competition for water resources across sectors and regions, the river basin has been recognized as the appropriate unit of analysis for addressing the challenges of water management. Modeling at river basin scale can provide essential information for policymakers in resource allocation decisions. A river basin system is made up of water sources components, in-stream and off-stream demand components, and intermediate (treatment and recycling) components. Thus a river basin is not only characterized by natural and physical processes but also by man-made projects and management policies. The essential relations within each component and the interrelations among these components in the basin can be represented in an integrated modeling framework. Integrated hydrologic and economic models are well equipped to assess water management and policy issues in a river basin setting. Some models of natural and physical processes were developed by some scholars, and were widely used in river basin management, such as BASIN, SWAT and MIKE etc. There are very few models to analyze the water use in social-economic process. Therefore, this work describes the methodology and application of an integrated hydrologic–economic river basin model.

Today we are faced largely with a "mature" water economy, and most research is conducted and focused on their demand to cope with an explicit recognition of resource limits (Cai et al. 2006). There has been much research focused on a multitude of situations that might be presented in a mature water economy. Much work has been motivated by expanding municipal and industrial demands within a context of static or more slowly growing agricultural demand. More recently, expanding interior demands of natural resource have been accelerated. On the supply side, conjunctive use of groundwater has been considered in addition to simple limits on surface water availability. In addition, some researchers have worked on waterlogging and water quality effects (primarily salinity). In this work, it examines a "complex" water economy: one in which demands grow differentially not only within but also among sectors, and one in which limited opportunities for increasing consumptive use exist. In particular, the growth of high-value

X. Deng et al., *Integrated River Basin Management*,
SpringerBriefs in Environmental Science,
DOI: 10.1007/978-3-662-43466-6_1, © The Author(s) 2014

irrigated crop production within the case study basin (the Heihe River Basin in China), together with the rapidly growing urban area, provides a rich context in which to examine the general problem of basin-level water management. At the same time, long-term aridity of nature has made the eco-environment in inland river basin located in northwest China much vulnerable, and immethodical exploitation and utilization of water resources has further deteriorated the situation.

The methodology presented is optimization with embedded simulation. Basin wide simulation of flow and water balances and crop growth are embedded with the optimization of water allocation, reservoir operation, and irrigation scheduling. The modeling framework is developed based on a river basin network, including multiple source nodes (reservoirs, aquifers, river reaches, and so on) and multiple demand sites along the river, including consumptive use locations for agricultural, municipal and industrial, and in-stream water uses. Economic benefits associated with water use are evaluated for different demand-management instruments—including markets for tradable water rights—based on the production and benefit functions of water use in the agricultural and urban-industrial sectors. The modeling framework includes multiple techniques, such as hydrologic modeling, spatial econometrics, geographic information system (GIS), and large-scale systems optimization. While these techniques have been adapted in other studies, this work represents a new effort to integrate them for analyzing water use at the regional level.

1.2 Overview of Input–Output Table

Input–output analysis is an analytical framework to analyze economy, which is developed by Leontief in the late 1930s, and the name is given to recognize the Nobel Prize in Economic Science that he received in 1973 (Leontief 1936; Leontief 1941). Input–output analysis is essentially to put forward a theory about the process of production, which is based on a particular type of production function. The main relationships are involving quantities of inputs and outputs in productive processes. Input–output framework fundamental purpose is to analyze the industry in the economic interdependence. Economic analysis of the basic concepts of a core component, but also the most widely used methods in economics (Baumol 2000).

To generally establish the basic input–output analysis according to the observation of a specific geographical area economic data, it is concerned about the activities of a group of industries, and depicts of the products (output) of industry and consumer products (input) from other industries in the production process. In practice, it may consider about thousands of industries, which are various. The input–output model needs to include a lot of basic information, which is contained in input–output tables. The basic information used in input–output tables covered the basic economic flow from each industrial sector in the process of production, considered as a producer, to each of the sectors, itself and others, considered as consumers. The row of the table describes the output distribution of producers in the whole economy. The column shows the input components of the specific industry from other industries. Additional columns, labeled as final demand, record the sales by each sector to final markets for their production, such as personal consumption purchases and sales to federal government. For example, electricity is sold to industry in other sectors, while it is as an input to production (inter-industry trade), and residential consumers (final demand sales). The additional rows, labeled Value Added, account for the other (non-industrial) inputs to production, such as labor, depreciation of capital, indirect business taxes, and imports.

The worldwide input–output analysis is developed to study the interdependence among many various different sectors in any economy (Miller and Blair 2009). An input–output table records the flow of products, in which each industry sector is both producer and consumers to other industries (Miller and Blair 2009). The input–output table presents a quite complete picture of economy on some particular point time, providing a series of important macroeconomic aggregates (production, demand components, value added and trade flows) and decompositions among the different industries and products. In addition, on one hand, the input–output table is a suitable instrument to perform structural analysis of the correspondent economy, depicting the interdependence between its different sectors and between the economy and outsides. On the other hand, the input–output table provides an important database to the construction of economy in input–output models which may be used, for example, to evaluate the economic impact caused by exogenous changes in final demand (Miller 1998).

Input–output table is a powerful tool presenting a very simple and basic concept, which is based on the output consumed by the factors of production (input) which can be, in their turn, the output of other industries. It can conduct economic analysis at any geographical level, such as local, regional, national and even international. There are a lot of developed input–output models so far. Input–output model can be used in a wide range of economic analyses. Firstly, it can describe and measure the composition and level of economic activity. Secondly, it can be applied to the impacts of changes in supply and demand of entire economy and analysis of the flows of goods and services between the industries and final consumers which help us understand the relationships among industries. In addition, it provides basic measurements and calculations of Gross Domestic Product (GDP). Input and output has also been extended to be a comprehensive framework for employment and social accounting metrics associated with industrial production and other economic activity, as well as to accommodate more specific explicitly issues such as international and inter-regional flows of goods and services, or activities of accounting and related inter-industry energy consumption and environmental pollution.

1.2.1 Theoretical Exploitation and Empirical Studies

The input–output analysis was developed by Leontief (1936), and to honor his works at interdependence research among industries and commodities production in economy structure he received the Nobel Prize in Economic Science in 1973. A general input–output table contains the valuable information about the market allocation of resources in an economic system. Institute for Prospective Technological Studies (IPTS) European Commission (1998) developed the input–output table with three major blocs, the intermediate consumption matrix, the final demand matrix and the primary inputs matrix. Ethiopian Development Research Institute (2009) developed the input–output table, in which rows describe the distribution of a producer's outputs throughout the economy, while the columns describe the composition of inputs required by a particular industry to produce its outputs.

The input–output table presents an economy as network of flows or linkages between economic activities specified as distinct sectors. It provides the underlying core database for a number of economic models which rely on restrictive assumptions that need to be tested before application. There are an increasing number of sophisticated models going to be used for assessing the impacts of economic change on other entities at the multiple levels. It is also be used to assess the distributional effects of change across the industries and regions included in the input–output table. If linked to household consumption and income data, the distributional effects of economic policy change on households can also be assessed. Now the input–output table is widely used at regional and national levels to analyze the influence of economic change on natural resources and environmental issues.

Regional input–output tables can be broadly divided into two types, regional input–output tables and inter-regional input–output tables. Regional input–output tables describe transactions in goods and services in a specific region during a given period. In contrast, inter-regional input–output tables cover multiple regions at the same time, describing transaction relationships of goods and services not only inside a particular region but among them as well. Since Isard (1951) developed the inter-regional input–output model (IRIO), other studies of input–output table working at inter-regional level come out successively. For instance, both Chenery et al. (1953) and Moses (1955) established the multiregional input–output (MRIO) model. Thereafter, a very thorough and detailed theory producing an inter-regional input–output table is provided by Isard and Langford (1971). Some early regional input–output models can be found in the studies of Polenske (1980) and in Geoffrey (1984).

Rich experience of research and application in regional input–output model are accumulated at abroad. Japan got the most obvious achievement in studying input–output model among countries. Institute of Developing Economies (IDE) developed an input–output model among countries including six worldwide regions: the United States, Europe, Oceania, Latin America, Asia, and Japan. Since then IDE tried to develop the input–output model among Asian countries. The attempt of IDE drove other countries including China trying to study the international input–output model (Zhang and Zhao 2006). Among the researches, the mainstream topic is the inter-regional trade flow estimation. Reed (1967) has developed the inflow and outflow models by using the railway and highway transportation data to analyze trade flows among different regions of India. Chisholm and O'Sullivan (1973) has used the gravity model consisted of 18 regions and 13 commodities to estimate the inter-regional trade in Britain. Graytak (1970) has drawn a conclusion that the feedback effect is important to developing the regional input–output table. Dietzenbacher (2002) has analyzed the relationships between spillover and feedback effect by using inter-regional input–output model. In recent decade, the inter-regional input–output model is commonly applied for carbon emission, because it is a powerful tool to analyze potential environmental pollution and gas emission that embedded in international trade among different countries (Wiedmann et al. 2007). Moreover, since the data barriers are eliminated with the development of international trade models such as GTAP, more researches are engaged in distinguishing the responsibility of gas emission due to the inter-regional and international trade by using MRIO models (Peters and Hertwich 2006; Wiedmann et al. 2008).

In China, many scholars began to study and try to compile the inter-regional input–output model in the late 1980s. Chen et al. (1988) developed a regional input–output model including two areas, North Jiangsu and South Jiangsu, by using the typical survey method. Akita et al. (1999) developed an inter-regional input–output model in northeast China and other parts of China through the location quotient method. Liu and Okamoto (2002) used non-survey method to construct the inter-regional input–output model in three major areas and 10 sectors of China under the framework of Leonief-Strout gravity model to estimate the inter-regional flow of China. Ichimura and Wang (2003) are the professors in Development Research Center of the State Council. They wrote a book, *Inter-regional Input–output Analysis of the Chinese Economy*, which systematically expounded the evaluation problems between methodology and data sources for compiling inter-regional input–output table. They also compiled the regional input–output tables of seven major areas of China which was used for policy analysis. State Information Center (2005) adopted multi-regional input–output (MRIO) model method to compile 1997 inter-regional input–output table of China. Besides, Research Center on Fictitious Economy and Data Science (Zhang and Shi 2011) created the inter-regional input–output model including 30 provinces based on the framework of Chenery-Moses model and the 2002 input–output table including 30 provinces of China, in which both the actual data and non-survey-based method were used. Liang (2007) separated China into eight economic regions through a multi-regional input–output model for energy requirements and CO_2 emissions in China to perform scenario and analyze sensitivity for each region in the years of both 2010 and 2020.

Since the conflicts between further economic development and natural resources depletion increasingly becoming severe, there is clear need to conduct researches to address the related issues in which the natural resources exploitation and consumption especially land and water resources are the most attractive concerns. The first systematic studies of the integrated input–output tables are developed by Statistics Netherlands under the name National Accounting Matrix including Environmental Accounts (NAMEA) at the end of the 1980s. Meanwhile, several EU Member States and supranational agencies started to collect related data in the 1980s and the 1990s, since then Eurostat is standardizing the NAMEA data gathering at national level, on a voluntary basis. EU inaugurated the input–output analysis integrated with environment information. This is followed by UN who formulated an international standard in 2003.

Regional input–output table plays a vital important role in revealing connections among sectors (products) or the economic and technological linkages among regions. It also develops the regional economic forecast in the middle or long term as well as distributes the productive factors reasonably. The county level input–output table can support to quantitative analysis of major proportional contribution of each sector, and further providing a logical basis for the strategies of generation. The sectorial composition of input–output table at county level regards the commodity production in each sector as output and relative major industrial production in each sector as input. Therefore the input–output table at county level has a combination of both commodity and industry features. As a regional input–output model at county level with more exquisite industries and sectors, it provides specific industrial sectorial output changes by evaluation of local economic policy, especially for certain lump sum policies implemented in various industries at the county level. Moreover, it can also provide structural demonstration of economic forecasting.

Generally, inchoate applications of the input–output model were carried out at national levels to assess interdependence of economic impacts through industries and sectors. Nowadays, input–output table is routinely applied in national economic analysis by specific institutions at different levels, such as state, industry, and the research community. In summary, input–output table can be used in many aspects in terms of enterprise management, macroeconomic analysis, policy simulation, economic forecast and environmental protection.

1.2.2 The Differences Between National and Regional Input–Output Study

The original application of input–output model is at national level. Nowadays, researchers are more interested in economic analysis at regional level, such as a group of the state, an individual status, a county, or a metropolitan area, because a modified input–output model would offer a policy-making perspective for specific problems in a region (administrative or geographical sub-national regions, such as a county or a basin). There are various basic input–output characteristics of a small regional economy different from national economy. According to the study of Miller and Blair (2009), there are two specific distinct characteristics to make necessary and evident distinction between national and regional input–output tables.

Firstly, the production structure of each region is specific and may be very different from countries, so the structure of input and output built at a specific area should be appropriate. Although in the output table, the input data of a state should be the average value of data of many specific areas, maybe it is the same in a small specific area, and it also can be significantly different from the input–output table of country. As to the early approaches given in regional input–output applications, there are some minor changes compared with countries' method, the use of the national input coefficients in the region has given way to coefficients tables that are tailored to a particular region on the basis of data specific to that region. Secondly, the regional input–output table focus on the economies of smaller scale, dependents more on the outside world (this including the other regions of the same country and other countries), and in which exports and imports become more important in the decision of demand and supply in the region. Since the 1950s, different input–output models have been developed, being distinguished through the criteria, such as the number of regions, the inter-regional linkages with other bordering regions, the degree of implicit detail in inter-regional trade flows and the kind of hypotheses assumed to estimate trade coefficients.

According to the criteria of regional input–output table mentioned above, the method of regional input–output analysis is different from the nationwide level. The first criterion is associated with several types of model to design a system with more than one region, which would distinguish single-region model from others. The single-region seeks to capture the impact of the region alone. Therefore, the fact of these single-region models' key limitation is that it ignores the impact link of one region with other regions. In reality, when one region increases its production, as a reaction to some exogenous change in its final demand, for example, some of the inputs that are needed to answer the production augment will come from the remaining regions, originating an increase of production in these regions, these are the spillover effects. The remaining region, in turn, may need to import the input from other regions (which may include a first region) to use in their production. The inter-regional feedback effects which are caused by the first region itself, and through the interactions the first region performs with the remaining regions (Miller 1998). The input–output analyses' application systems and multiple regions are fully considering the impact of interconnection in different regions (equivalent to the second criterion mentioned above), it is a significant contribution of Walter Isard (Glasmeier 2004). These contributions originated inter-regional model, which is also known as Isard model. The practical difficulties of implementing the regional model mainly due to its high demand of inter-regional trade data, and encourage the emergence of multi-regional model (the model of Chenery-Moses is the most popular).

1.2.3 The User Needs of Compiling Regional Input–Output Tables

A brief review of the regional input–output analysis makes it clear that their implementation requires access to some data on interregional trade flows. Many regional studies have demonstrated that a regional trade flows between the remaining areas with other regions are more significant than the trade flows between these regions with foreign countries (Munroe et al. 2007). In addition, growth rate in intra-regional trade is indeed faster and more frequent than the trade between regional and international (Jackson and Murray 2004). One of the reasons of the rapid growth of trade is that it is to replace the area transactions, this process is called "hollowing out" of the facts: it means the relationship of economic density tends to decrease between the regions in favor of inter-regional contact (Polenske and Hewings 2003). Given the relative importance of trade flows of its intra-regional trade, so the knowledge of the volume and nature of inter-regional trade flows constitutes a critical issue for regional analysis.

When the deficit appears in the trade balance of a region, it means that the region relies on income transfers and grants savings from other adjacent regions or from elsewhere outside of the region in the world (Sargento and Ramos 2003). In a more detailed perspective, recognized about the regional external trade, segmented by commodities, allows us to characterize productive specialization, foresee eventual productive weaknesses as well as determine the region's dependency on the exterior (or in some cases the exterior's dependency on the region) regarding to the supply of different commodities. However, it is hard to establish inter-regional trade flows among regions to find the data to implement the input–output model. At the regional level, the input–output model is different from the original. It can be a broader economic base model on the one hand and the less complex econometric model visualization on the other hand. Some linkages of input–output analysis with other standard modeling techniques will be presented and further denoted that the input–output table has a considerable degree of flexibility in its contribution to a good understanding of the structure of an economy will be reinforced. It is really necessary to the preparation of input–output tables in a small region to clarify inter-regional trade flows. Besides, China has a long tradition in compiling input–output tables at the regional level, next to the compilation of national input–output tables. Except for Tibet, for 30 out of 31 provinces survey-based input–output tables have been constructed every five years since 1987 by the regional statistical bureaus following the compilation scheme for the national input–output tables (Qi 2007).

1.3 Overview of Computable General Equilibrium (CGE) Model

1.3.1 Application of CGE Model in Water Management

The CGE models address water related issues at national/regional scale precedes the analysis at global scale. One of the first efforts in this domain was done by Lofting and Mcgauhey (1968), in which they include water in an input–output model in order to analyze the requirements of water in California. Since then, CGE model has been applied for analyzing a broad range of issues, such as, water pricing policy (Decaluwe et al. 1999; Letsoalo et al. 2007), water quality (Deng et al. 2010), water allocation (Seung et al. 2000; Diao et al. 2005; Diukanova 2011), water markets (Gomez et al. 2004), irrigation policies (Elsaeed 2012) and climate change impact (Cai et al. 2008; Dudu and Cakmak 2011) and so on.

Explicitly, Decaluwe et al. (1999) present a CGE model applied to Morocco in which they analyze the impacts of different pricing policies on water allocation. The pricing scenarios analyzed are: Boiteux-Ramsey pricing (BRP), BRP and tax decrease, BRP and income tax decrease marginal cost pricing (MCP), and MCP with tax decrease. The model presents detailed input–output changes of agricultural sector through a series of nested CES function depicting interdependence among outputs of agricultural production, fertilizer, water, land, capital and labor by their respective inputs. The intermediated consumption is represented through a composite. The model considers two different technologies to produce water: water storage by build-up dams would be theoretically substituted for capital production and water "produced" efficiently by both circulative surface water and pumped water from underground. Letsoalo et al. (2007) indicated the setup of water charge generates triple dividends for South African economy. In that case, the potential reduction in water usage is considered as the first dividend spending those revenues for stimulating economic growth would bring about the second dividend back to economy system, hence the allocation of household income would be redistributed by higher level of economic growth which generates the third dividend to social welfare improvement.

Regarding water allocation, Seung et al. (2000) constructed a dynamic CGE model to analyze the economic impacts of water reallocation on outputs of agriculture and other main sectors with local increasing recreational demand of land use. The model considers 8 aggregated sectors, 3 of which are about agriculture production which are close related to natural resource utilization. Thus, land use would be a critical tangent point to analyze the relationship between economy structure and natural resource utilization. For all outputs in each sector presented by a Cob-Douglas function. For instance, the agricultural sector outputs are produced by how much land use, capital and labor inputs, and technology coefficient assumed as a constant. In this model water is entitled as the price of water rights associated to the relative amount of land property. They supposed a reduction of land use for agricultural sectors if water is extracted from the land.

Diao and Roe (2003) presented a CGE model analyzing the impact of a trade liberalization policy on water resource allocation. They proposed a theoretical model which links trade reform with water market creation. According to this model, the combination of a trade reform with the creation of a water rights market generates the most efficient allocation of water in economic perspective. He and his colleagues then present a CGE model analyzing the economy-wide effects of water reallocation to figure out its most productive usage. The model differentiates also between irrigated and rainfed crops. An extension of the model was built in 2008 and the new version of the model includes explicitly a difference between surface and groundwater (Diao et al. 2008).

The agricultural sector is modeled using a series of nested CES function for the primary inputs, while the intermediate consumption is assumed to be Leontief function. Agricultural production use labor, capital, land and water. Labor could be rural or urban. Rural labor is mobile only among agricultural sectors. Capital and land are mobile within irrigation zones. Land could be irrigated or rainfed, and the supply of irrigated land is fixed. Finally, water is mobile within each region but not across regions.

Lennox and Diukanova (2011) present a regional CGE model suitable for the analysis of water policies. In their modeling approach, the agricultural sectors are presented through a series of nested CES functions for the inputs, in which the agricultural production uses labor and composite land and capital. The composite land and capital is further disaggregated into the demand of land and the demand of capital. At the bottom of the productive structure, water is linked in fixed proportion with the land endowment. Juana et al. (2011) analyzed the economic impact of the reallocation of water from the agricultural sector to other sectors on the South African economy. Using information from 19 water management areas, they define the amount of water used by each sector, while using the municipal water tariff schedule they assign the monetary value of water used by sector. The model considers water as a new primary factor. Along with capital and labor, the production structure is modeled using CES functions with the exception of capital that it is modeled through fixed proportions. Water and labor are freely mobile across sectors, while capital is sector specific. And based on the GTAP-E model (Burniaux and Truong 2002), using the aggregation of GTAP 5 database (based on 1997), Berritella et al. (2007) propose a new modeling approach called GTAP-W that explicitly considers water as a production factor. Using the Leontief formulation, water is combined with the value-added energy nested and intermediate input at the top of the production tree. This formulation implies no substitution among these three components, thus water cannot be substituted with any other input. Calzadilla et al. (2010) present a new CGE model addressing water related issues. This model presents a major improvement in contrast to previous version that the new version considers the difference between water provision systems, such as rainfall and irrigation. The model is considered using indirect approach, differentiating between rainfed and irrigated crops. The new approach consists of splitting the original land endowment, in the value-added nest, into 3 components: pasture land, rainfed land and irrigated land.

1.3.2 An Integrated CGE Model with Resource and Environment Account

Computable General Equilibrium (CGE) models are widely used in policy analysis, tracking resource flow, and analyses of environmental issues. For environmental policies that are expected to affect many sectors either through direct compliance costs or indirectly through linkages between sectors of the economy (i.e., industries, households, government, trade), it may be important to account for these interactions and constraints. General Equilibrium (GE) models account for these linkages and are more appropriate than Partial Equilibrium (PE) analysis of large regulations that are expected to have measurable impacts across the economy. This work describes the CGE version of economic model for water resource policy analysis, which was specifically designed for analyzing large-scale water management.

As to the platform for CGE simulation, GEMPACK, a flexible system for solving CGE models, is often used for formulating and solving CGE model through the percentage-change approach. GEMPACK automates the process of translating the model specification into a model solution program. The GEMPACK users need no programming skills, instead, they just need to create a text file, list the equations of the model. The syntax of this file resembles ordinary algebraic notation. The GEMPACK program, TABLO, then translates this text file into a model-specific program which solves the model.

With rapidly growing populations, many countries in the world have found it is difficult to meet municipal, agricultural, and environmental water demands simultaneously. Moreover, water resources are faced with several stresses of quantity and quality in inland river basin, which are closely intervened by the human activities in the fields such as agriculture, industry, land use/cover change, and climate change, and so forth. Considering the critical role that water plays in agricultural production, any shock in water availability will have great implications on agricultural production, and through agricultural markets these impacts will reach the whole economy with economy-wide consequences. The relationship between water resources and economy structure is complicated in regional extended-IO table. Therefore, this complexity motivates the need for analytical methods, which can take interrelated markets and secondary impacts on evaluating the net effects from changes that affect water resources into account.

Some researchers think the economic consequences of water management policies are different in different districts of basin. At the same time, a long-term plan on water resources is also necessary to sustainable development of economy and environment. Therefore, the multiregional and dynamical analysis model is suitable to validate these deductions.

In this work, a new modeling approach is explicitly presented which aims to embed water and land resource into the CGE model. In order to reach this objective, a model framework has to be built by considering the water endowment of water use efficiency and water price for agriculture and other industries, water and land resource allocation and unitary irrigation costs. Therefore, we will introduce these contents as follows.

Water is a vital resource in any economy and at all stages of economic development. In the past years, people were located in close proximity to waterways that not only convenient to transportation, but also vitally important to cultivated land irrigation of crops and feeding livestock for human consumption. Human's growing food demand for water storage, flood control constructions, irrigation activities, and power generation projects resulted in building infrastructures like dams and reservoirs which added the dimension of controlling the flow and timing to the value of water resources. As economies become more complex, so do the multiple demands for water resources. In the consideration of the "value of water", the entire spectrum of consumptive and non-consumptive water uses need to be considered, as these varied uses are often integrated. This spectrum includes various combinations of water attributes: timing of water flows; quantity of water; and physical quality of water. The demand for water for a single given use could be based on specific minimum requirements for all of these attributes. And there are many different uses for water.

A single water resource could be used to satisfy many different demands for water within an economy. For example, water within a river system is generally limited to being used within that river's watershed, but not limited to any one use. Some water usages do not consume water resource; hydroelectric generation is one such example of a non-consumptive use. Flood irrigation of croplands is an example of partial consumption of water, to move the water across the field, a quantity is diverted sufficient to create a flow that will cover the field, but only water that is evaporated or transpired is consumed. The next user downstream has access to balance of water flows, including returns from irrigation systems (return flows). A municipal water utility could divert water for the production of drinking water for the people. This utility is producing at least two new products from the raw water: drinking (potable) water; and conveyance (delivery) of potable water to living place. This water can be used for direct consumption by individuals (residential), used as an input for a business, or used as a source of irrigation for community landscaping. All of these water usages are limited to the watershed, unless an infrastructure is constructed to allow for the export of water from one basin to another.

In the above examples, the same water resource is used at different times and in different ways with bringing diverse values to an economy, either by producing goods and services or as a final demand product. Each of them has different value associated with the water usage from others did. Thus it is not really possible to state any single value for water, but rather a water resource used within a watershed (or economy) results in a total increase of economic value or human welfare. A general equilibrium approach to understanding an entire regional economy through its industrial sectors and market inter-linkages presents a reasonable way of valuing water resources. CGE models are simulations based on general equilibrium theory. CGE models allow for the multiple types of water usages resource in an economy and return estimates of changes in social welfare for increasing or decreasing in the water resource, thus providing an estimate of the marginal value of water to the economy.

The well-defined functional market would support commodities trade-off balance between prices and quantities in an equilibrium system. In that 'ideal' pure market economy, water could be treated as normal goods, and the value of water would equal its marginal value product (MVP) with the quantity of water purchased at that price by firms in each sector. In other words, MVP would depend only on each sector's demands for water through competitive market based on both supply and demand sides in the market clear equilibrium of water price and quantity. However, this ideal world does not exist. Water is normal goods. Water as one of natural resource highly depends on geographical characteristics in a region. 'Prices' associated with water usage do not necessarily reflect the marginal value of water, but rather may be administratively set and subject to governmental subsidies, regulations, or restrictions of water rights in a certain institution. In part, this is because several factors make it difficult to specify exclusive private property rights for water.

1.3.3 Challenges Facing the Integrated CGE Models

Building the integrated CGE model is an extremely data intensive enterprise, requiring detailed baseline data for all parts of the economy. Building into a CGE model the ability to address changes in water resources usually requires additional data that links economic sectors with their water use. Wittwer and Griffith (2012) refer to this data as "water accounts". Finding sources for water accounts data can be a challenge. The integrated CGE model represents a minimum level of water data, i.e. estimates of water use per CNY of output by sector, household water use, and total water use for the regional economy as a whole. For many policy questions, however, a greater effort collecting and organizing this data is necessary. The development and availability of water accounts data is the major reason why Morocco, South Africa and Australia have an abundance of such integrated CGE models. Unfortunately, some research regions face a lack of water accounts data in China. The existing integrated CGE models used a variety of unique datasets on water resources, allocation and usage.

CGE models assume a simplified version of a neoclassical economy that has a point of equilibrium where the right price creates market clearing. Special techniques are required when water is known as a resource with observed prices detected by a well-functional market. It is a challenge to estimate a starting 'market' price for the initial baseline equilibrium which is used to calibrate the model. It also requires different techniques to treat the water as a factor or a sector but both of them would confront that a well-functioning market would never generate market-clearing prices and quantities. We observe a variety of methods for tackling these difficulties:

- Estimation of water 'rent' in various ways, often use land values. The rent is then subtracted from gross operating surplus and distributed to households. This technique assumes a functioning water factor market (for example, Robinson and Gehlhar 1995; TERM-H2O models described in Dixon et al. 2012; Seung et al. 2000, etc.).
- Assumption that no market for water exists in the baseline, that water is in surplus and its price in equilibrium is zero which becomes positive only as water supplies are withdrawn (Diao et al. 2005).

- In one case, water 'rent' is subtracted from utility fees charged to industries (Hassan and Thurlow 2011).
- Some models use administratively set utility fees for treated water as if they were determined by market equilibrium.

The amount of water use by each industry sector typically for estimating a water intensity factor gives the amount of water which is necessary for producing a unit of output. Depending on the purposes of the model, water use by households and government, total water availability in the region, and other more detailed water data may be necessary. If a major focus of an integrated CGE model is to find an economic value for water, the boundaries may be defined as including all economic activities within the same hydrological basin or which is connected through pipelines. By definition, water from outside this boundary does not enter the regional economy. Outside of these boundaries, water supplied from different sources is simply not available for the target CGE economy. Therefore, the marginal value of water in separate economic systems, described by separate CGE models, would not be expected to be the same, because by definition there is no trade, and therefore market-clearing conditions is difficult to hold.

Region-specific demands for water resources also vary widely depending on annual precipitation, average temperature, industry mix and technologies, and consumer preferences. Region-specific shocks to water resource supply and demand can also vary immensely by region. Given differences between regions, many water policy questions among economically linked regions can be better answered with a multi-regional model that incorporates regions closely aligned to the watershed level models that investigate water trading among regions.

References

Akita, T., Xie, B., & Kawamura, K. (1999). The regional economic development of Northeast China: An inter-regional input–output analysis. *JESNA, 1*(1), 53–78.

Baumol, W. J. (2000). Leontief's great leap forward: Beyond Quesnay, Marx and von Bortkiewicz. *Economic Systems Research, 12*, 141–152. doi:10.1080/09535310050005662

Berritella, M., Hoekstra, A. Y., Rehdanz, K., Roson, R., & Tol, R. S. J. (2007). The economic impact of restricted water supply: A computable general equilibrium analysis. *Water Research, 41*, 1799–1813.

Burniaux, J. M., & Truong, T. P. (2002). *GTAP-E: An energy-environmental version of the GTAP model*. GTAP Technical Papers. https://www.gtap.agecon.purdue.edu/resources/download/1203.pdf

Cai, X., Ringler, C., & Rosegrant, M. W. (2006). Modelling water resources management at the basin level: Methodology and application to the Maipo River basin. http://www.ifpri.org/sites/default/files/publications/rr149.pdf

Cai, X., Ringler, C., & You, J. Y. (2008). Substitution between water and other agricultural inputs: Implications for water conservation in a river basin context. *Ecological Economics, 66*(1), 38–50.

Calzadilla, A., Rehdanz, K., & Tol, R. S. J. (2010). The economic impact of more sustainable water use in agriculture: A computable general equilibrium analysis. *Journal of Hydrology, 384*(3), 292–305.

Chen, X., Shao, H., & Li, L. (1988). *The theory and practice of input–output in contemporary China*. Beijing: China Radio International press.

Chenery, H., Clark, P., & Pinna, V. (1953). *Regional analysis in the structure and growth of the Italian economy*. Rome: U.S. Mutual Security Agency.

Chisholm, M., & O'Sullivan, P. (1973). *Freight flows and spatial aspects of the British economy*. New York: Cambridge University Press.

Decaluwe, B., Patry, A., & Savard, L. (1999). *When water is no longer heaven sent: Comparative pricing analyzing in a CGE model*. Working Paper 9908. CRÉFA 99-05, University of Laval.

Deng, X., Zhao, Y., Wu, F., Lin, Y., Lu, Q., & Dai, J. (2010) Analysis of the trade-off between economic growth and the reduction of nitrogen and phosphorus emissions in the Poyang Lake Watershed, China. *Ecological Modelling, 222*(2), 330–336. doi:http://dx.doi.org/10.1016/j.ecolmodel.2010.08.032

Deng, X. (2011). *Modeling the dynamics and consequences of land system change*. Berlin: Springer.

Diao, X., Dinar, A., Roe, T., Tsur, Y. (2008). A general equilibrium analysis of conjunctive ground and surface water use with an application to Morocco. *Agricultural Economics, 38*(2), 117–135. doi:10.1111/j.1574-0862.2008.00287.x

Diao, X., Roe, T. (2003). Can a water market avert the "double-whammy" of trade reform and lead to a "win–win" outcome? *Journal of Environmental Economics and Management, 45*(3), 708–723. doi:http://dx.doi.org/10.1016/S0095-0696(02)00019-0

Diao, X., Roe, T., Doukkali, R., Doukkali, R. (2005). Economy-wide gains from decentralized water allocation in a spatially heterogenous agricultural economy. *Environment and Development Economics, 10*(3), 249–269. doi:http://dx.doi.org/10.1017/S1355770X05002068

Dietzenbacher, E. (2002). Inter-regional multipliers: Looking backward, looking forward. *Regional Studies, 36*(2), 125–136. doi:10.1080/00343400220121918

Dietzenbacher, E., & Temurshoev, U. (2012). Input–output impact analysis in current or constant prices: Does it matter? *Journal of Economic Structures, 1*(1), 1–18. doi:10.1186/2193-2409-1-4

Diukanova, O. (2011). Modelling regional general equilibrium effects and irrigation in Canterbury. *Water Policy, 13*(2), 250–264. doi:10.2166/wp.2010.090

Dixon, P. B., Rimmer, M. T., & Wittwer, G. (2012). The theory of TERM-H2O. In G. Wittwer (Ed.), *Economic modeling of water*. Dordrecht: Springer.

Dudu, H., & Cakmak, E. H. (2011). Regional impact of the climate change: A CGE analysis for Turkey. In *Proceedings of Politics and Economic Development: ERF 17th Annual Conference, Turkey ERF*.

Elsaeed, G. (2012). *Effects of climate change on Egypt's water supply. National security and human health implications of climate change*. Dordrecht: Springer.

Geoffrey, J. D. (1984). The role of prior information in updating regional input–output models. *Socio-Economic Planning Sciences, 18*(5), 319–336.

Glasmeier, A. (2004) Geographic intersections of regional science: Reflections on Walter Izard's contributions to geography. *Journal of Geographical Systems, 6*(1), 27–41. doi:10.1007/s10109-003-0121-0

Gomez, C. M., Tirado, D., & Rey-Maquieira, J. (2004). Water exchanges versus water works: Insights from a computable general equilibrium model for the Balearic Islands. *Water Resources Research, 40*(10), W10502. doi:10.1029/2004WR003235

Graytak, D. (1970). Regional impact of inter-regional trade in input–output analysis. *Papers of the Regional Science Association, 25*(1), 203–217. doi:10.1111/j.1435-5597.1970.tb01486.x

Hassan, R., & Thurlow, J. (2011). Macro–micro feedback links of water management in South Africa: CGE analyses of selected policy regimes. *Agricultural Economics, 42*(2), 235–247. doi:10.1111/j.1574-0862.2010.00511.x

Ichimura, S., Wang, H. (2003). *Inter-regional input–output analysis of the Chinese economy*. Singapore: World Scientific.

Isard, W. (1951). International and regional input–output analysis: A model of a space economy. *The Review of Economics and Statistics, 33*(4), 318–328.

Isard, W., & Langford, T. W. (1971). *Regional input–output study: Recollections, reflections, and diverse notes on the Philadelphia experience.* Cambridge, MA: MIT Press.

Jackson, R., & Murray, A. (2004). Alternative input–output matrix updating formulations. *Economic Systems Research, 16*(2), 135–148. doi:10.1080/0953531042000219268

Juana, J. S., Strzepek, K. M., & Kirsten, J. F. (2011). Market efficiency and welfare effects of inter-sectoral water allocation in South Africa. *Water Policy, 13*(2), 220–231. doi:10.2166/wp.2010.096

Leontief, W. W. (1936). Quantitative input and output relations in the economic systems of the United States. *The Review of Economics and Statistics, 18*, 105–125. doi:10.2307/1927837

Leontief, W. W. (1941). *The structure of American economy 1919–1929.* Cambridge, MA: Harvard University Press. doi:10.2307/1805821

Lennox, J. A., & Diukanova, O. (2011). Modelling regional general equilibrium effects and irrigation in Canterbury. *Water Policy, 13*(2), 250–264. doi:10.2166/wp.2010.090

Letsoalo, A., Blignaut, J., De Wet, T., De Wit, M., Hess, S., & Richard, S. J. T., et al. (2007). Triple dividends of water consumption charges in South Africa. *Water Resources Research, 43*(5), W05412. doi:10.1029/2005WR004076

Liang, Q. M. (2007). Multi-regional input–output model for regional energy requirements and CO2 emissions in China. *Energy Policy, 35*, 1685–1700.

Liu, Q., & Okamoto, M. (2002) The compilation and problem of China's regional input–output model. *Statistical Research, 9*, 58–64. [in Chinese].

Lofting, E. M., & Mcgauhey, P. H. (1968). Economic valuation of water. An input–output analysis of California water requirements. In *Contribution*, (Vol. 116). Berkeley: Water Resources Center.

Miller, R. E. (1998). Regional and inter-regional input–output analysis. In W. Isard et al. (Ed.) *Methods of inter-regional and regional analysis* (pp. 41–134). Brookfield: Ashgate.

Miller, R. E., & Blair, P. D. (2009). *Input–output analysis: foundations and extensions.* Cambridge: Cambridge University Press.

Moses, L. N. (1995). The stability of inter-regional trading patterns and input–output analysis. *American Economic Review, 45*, 803–826.

Munroe, D. K., Hewings, G. J., & Guo, D. (2007). The role of intraindustry trade in inter-regional trade in the Midwest of the US. In *Globalization and regional economic modeling*. Berlin: Springer.

Peters, G. P., & Hertwich, E. G. (2006). Structural analysis of international trade: Environmental impacts of Norway. *Economic Systems Research, 18*(2), 155–181. doi:10.1080/09535310600653008

Polenske, K. R. (1980). The U.S. multiregional input–output accounts and model D.C. Heath and Company Lexington, MA, USA. http://ideas.repec.org/a/eee/jpolmo/v4y1982i2p275-275.Html

Polenske, K. R., & Hewings, G. J. (2003). Trade and spatial economic interdependence. *Papers in Regional Science, 83*(1), 269–289. doi:10.1007/s10110-003-0186-7

Qi, S. (2007) *Introduction to the compiling system of the Chinese regional input–output tables.* Paper presented at the seventh Chinese input–output conference, Nanjing, China [in Chinese].

Reed, W. E. (1967). *Areal interaction in Indian: Commodity flows in the Bengal–Bihar industrial area.* Research Papers Series No. 110 Department of Geography, the University of Chicago.

Robinson, S., & Gehlhar, C. (1995). Land, water, and agriculture in Egypt: The economywide impact of policy reform. http://www.ifpri.org/sites/default/files/publications/tmdp01.pdf

Sargento, A. L. M., & Ramos, P. M. N. (2003). Estimating trade flows between Portuguese regions using an input–output approach. In *ERSA conference papers (No. ersa03p118) European Regional Science Association.* http://ideas.repec.org/p/wiw/wiwrsa/ersa03p118.html

Seung, C., Harris, T., Englin, J., & Netusil, N. (2000). Application of a computable general equilibrium (CGE) models to evaluate surface water relocation policies. *The Review of Regional Studies, 29*(2), 139–156.

State Information Center. (2005). *China's regional input–output tables*. Beijing: Social Sciences Academic Press.

Wiedmann, T., Lenzen, M., & Turner, K. (2007). Examining the global environmental impact of regional consumption activities. Part 2: Review of input–output models for the assessment of environmental impacts embodied in trade. *Ecological Economics, 61*(1), 15–26.

Wiedmann, T., Wood, R., Lenzen, M., Minx, J., Guan, D., & Barrett, J. (2008). Development of an embedded carbon emissions indicator—producing a time series of input–output tables and embedded carbon dioxide emissions for the UK by using a MRIO data optimisation system. Final Report to the Department for Environment, Food and Rural Affairs by Stockholm Environment Institute at the University of York and Centre for Integrated Sustainability Analysis at the University of Sydney. Project Ref.: EV02033, July 2008, Defra, London, UK (2008). Executive summary: http://randd.defra.gov.uk/Document.aspx?Document= EV02033_7333_EXE.pdf

Wittwer, G., & Griffith, M. (2012). *The economic consequences of a prolonged drought in the Southern Murray-Darling Basin*. Dordrecht: Springer.

Zhang, Y. X., & Zhao, K. (2006). *Inter-regional input–output analysis*. Beijing: Social Sciences Academic Press. [in Chinese].

Zhang, Z. Y., & Shi, M. J. (2011). Intra-industrial trade and interregional structural isomorphism of manufacturing industry based on China-IRIO2002. *Acta Geographica Sinica, 66*(6), 732–740. [in Chinese]

Chapter 2
Approach of Input–Output Table at Regional Level

Input–output analysis involves all aspects of the national accounts related to goods and services, including expenditure aggregates. Input–output analysis provides the opportunity to reconcile supply and use of goods and services, as well as reconcile GDP and expenditure on GDP. One of the goals of this analysis is to eliminate the statistical discrepancy. This is also a requirement for deriving downstream input–output tables. Compiling regional input–output table, not only identify the quantity of products in inter-regional trade, but also determine the trade flows among departments. Moreover, in the inter-regional trade, it is also necessary to distinguish how much of intermediate inputs used in the production sector and how much used in final consumption. Therefore, compiling inter-regional input–output tables require high quality data, but so far, apart from a small part of developed countries, the vast majority of countries cannot meet the need of basic data requirements compiling inter-regional input–output tables in the existing statistical system, because of a lot of manpower and material resources to carry out surveys and collect data, which makes considerable difficult to compile inter-regional input–output table at present. It requires compiling inter-regional input–output tables when the data resources are relatively low.

2.1 Methods of Commodity Flows Estimation

The key step of compilation of regional input–output table through a regional input–output model is to estimate the flows of commodity. As the statistical system is not perfect, the majority of countries are difficult to obtain the data of commodity flow directly which could compile regional flow matrix. Many studies are focused on how to estimate the flow of commodity according to reliable mathematical models and existing data. At present, the countries who research on this aspect in the world more in-depth include the United States, Japan, Russia, Finland, Spain, etc. Here are three common methods of estimating regional commodity flow, including location quotients, gravity model and regression equation, respectively.

X. Deng et al., *Integrated River Basin Management*,
SpringerBriefs in Environmental Science,
DOI: 10.1007/978-3-662-43466-6_2, © The Author(s) 2014

2.1.1 Location Quotients

The widespread use of the Location Quotients (LQ) approach for constructing regional input–output tables is primarily driven by pragmatic concerns. Generally speaking, detailed data are seldom available at the regional level. It is typically beyond the means of the users to implement more accurate methods and collecting the primary data which is needed. One way to solve this problem is to draw on a published input–output table pertaining to a larger geography and use employment based location quotients to estimate a local sub-section of that table. Implicitly by going down that route, the researcher is accepting some rather bold assumptions. For these, Harris and Liu (1998) refer to Norcliffe (1983), who identifies the main assumptions underlying the use of location quotients, to identify the export base in export base models.

It is clear that in order to let a region's share of national employment accurately represent its share of national production, there must be identical productivity per employee in each region in each industry for employment to be used as a proxy. Also, for similar reasons, there must be identical consumption per employee. However, there must be no cross-hauling between regions of products belonging to the same industrial category, so as not to underestimate inter-regional trade. Because these assumptions rarely hold, a number of researchers have attempted firstly to estimate empirically the extent to which the breakdown of these assumptions will influence estimates for input–output accounts and then come up with modifications of the LQ approaches that might counter some of the biases.

Various LQ methods have been suggested in the literature (Miller and Blair 2009). In general LQ approaches adjust the national technical coefficient to take account of the potential for satisfying input needs locally. A regional Input–Output technical coefficient is a function of the location quotient and the national technical coefficient:

$$a_{ij}^{RR} = a_{ij}^{RR}\left(LQ^R, a_{ij}^N\right) \tag{2.1}$$

where a_{ij}^{RR} is the regional IO technical coefficient, LQ_i^R is the location quotient and a_{ij}^N is the national technical coefficient.

1. Simple location quotient (SLQ)
The simple location quotient for sector i in region R is defined as:

$$SLQ_i^R = \left[\frac{E_i^R/E^R}{E_i^N/E^N}\right] \tag{2.2}$$

where E_i^R and E^R are employment in sector i in region R and total employment in region R respectively, and E_i^N and E^N are employment in sector i and total employment in the nation as a whole.

When the SLQ_i^R is greater than one (less than one), it can be inferred that sector i is more (less) concentrated in region R than in the nation as a whole. Where the location quotient is less than one the region is perceived to be less able to satisfy regional demand for its output, and the national coefficients are adjusted downwards by multiplying them by the location quotient for sector i in region R. Where the sector is more concentrated in the region than the nation at large ($LQ_i > 1$), it is assumed that the regional sector has the same coefficients as the nation as a whole. Therefore for row i of the regional table:

$$a_{ij}^{RR} = \begin{cases} a_{ij}^N SLQ_i^R & \text{if } SLQ_i^R < 1 \\ a_{ij}^N & \text{if } SLQ_i^R \geq 1 \end{cases} \qquad (2.3)$$

2. Cross industry location quotient (CILQ)
A criticism of the simple location quotient is that it does not take into account the relative size of the sectors engaged in intermediate transactions. The argument goes that if a sector which is relatively small locally is supplying a sector which is relatively big, this should imply a need for imports to satisfy intermediate demand, and vice versa. This is addressed with cross industry location quotients (CILC). The CILQ for sectors i and j can be defined as:

$$CILQ_{ij}^R = \frac{SLQ_i^R}{SLQ_j^R} \left[\frac{E_i^R / E_i^N}{E_j^N / E_j^N} \right] \qquad (2.4)$$

where sector i is assumed to be supplying inputs to sector j. As with the SLQ national coefficients are not adjusted if $CILQ_{ij}^R \geq 1$ as it is assumed that intermediate demand can be met within the economy.

2.1.2 Gravity Model

In the development of regional input–output analysis, gravity model is used to calculate regional trade of industrial products and it is decided by the following formula:

$$t_i^{RS} = \frac{x_i^R d_i^S}{\sum_R x_i^R} Q_i^{RS} \qquad (2.5)$$

where t_i^{RS} represents the outflow volume in sector i from region R to region S, x_i^R is gross output (gross supply) in sector i in region R, d_i^S is the gross product demand from region S to sector i, $\sum_R x_i^R$ is the gross output (gross demand) of all sector i,

Q_i^{RS} is the trade coefficients in sector i from region R to region S, or the coefficient of friction.

The key factor of using the gravity model is to estimate the coefficient of friction. Lcontief and Strout put forward the corresponding estimation method on the basis of different data. Ihara (1979) introduced the proportional distribution coefficient of inter-regional commodity flows to calculate the trade friction coefficient of different products. The calculation method of the proportional distribution coefficient of inter-regional commodity flows assumes that there are similarities of the distribution proportion of commodity flows from one region to the other regions and the most important product allocation proportion. Thus the distribution coefficient can be treated as the regional product trade flow parameter, Q_i^{RS}, which can be defined as:

$$Q_i^{RS} = \frac{H_i^{RS}}{H_i^{RO} H_i^{OS} / H_i^{OO}} \tag{2.6}$$

where H_i^{RS} is the trade flow in sector i from region R to region S, H_i^{RO} is the amount of the products in sector i of region R, H_i^{OS} is the gross import in the region S, H_i^{OO} is the gross export of the products in all the sectors i. The larger the coefficient of regional product trade flows is, the closer linkage between the regions is.

2.1.3 Regression Equation

The regression equation has three kinds of common forms.

The cross section data equation:

$$T^{RS} = \alpha^{RS} IND^R + \beta^{RS} GDP^S + \delta^{RS} D^{RS} + e \tag{2.7}$$

Time series equation:

$$T^{RS} = \alpha^{RS} IND^R + \beta^{RS} GDP^S + \theta^{RS} t + e \tag{2.8}$$

Mixed time and space equation:

$$T^{RS} = \alpha^{RS} IND^R + \beta^{RS} GDP^S + \theta^{RS} t + \delta^{RS} D^{RS} + e \tag{2.9}$$

The dependent variables include the industrial gross output (*IND*), the sector output as the start point (region R), GDP as the end point (region S), D is the direct distance between provinces and t is the time, and the independent variable is the mixed production flows which are obtained from traffic yearbook. This equation reveals the fact that the regional product flows are decided by the ability of supply and demand of one region.

The main idea of regional flows estimated by regression equations is to measure the elasticity of dependent and independent variables by using the sample data. Thereafter, the gross output IND_i^R of sector i is introduced, then the flows T_i^{RS} in sector i from region R to region S can be obtained finally. The important assumption of the regression equation is that the elastic value is a set of parameters which can reflect the movement of fluid diffusion in a certain period of time. If the parameters of a certain period of time are estimated based on the data of mixed production, then the regional flows of the pure sector can be calculated by the parameters and relevant data. Finally the regional trade flows and the coefficient matrix of the corresponding region can be obtained.

2.2 Methods of Input–Output Table Compilation

In this work, the methods of constructing a regional input–output table can be roughly divided into two categories including survey-based method and non-survey-based method. One of the key problems of input–output tables is that the survey-based method is extremely time consuming and therefore expensive. Therefore high level input–output tables are compiled by the specific authorities. For example, the major job of Bureau of Economic Analysis is to compile the Benchmark input–output table for the U.S. The production cycle is now generally surveyed every five years, for years ending in '2' and '7', which is dictated by the schedule of the Economic Census. However, with the development of economy, sectors and industries are increasingly becoming more complex. This is illustrated by the history of input–output table of Japan. The first compilation of the input–output table for Japan could date back to 1955 as the reference year. Thereafter compiling input–output tables came to be a joint work by related ministries and agencies every five years. The 1951 input–output table is compiled by the Environmental Protection Agency (EPA) and by the Ministry of International Trade and Industry (MITI). From the experimentation phase to the stage of practical use, the 1955 input–output table has a higher accuracy than that in 1951. Moreover, there were remarkable changes in 1960 input–output table because of technical innovation from which they had to seek materials for reviewing input–output table as of doubling national income. The 1965 input–output table consisting of 456 rows and 339 columns is established and published as the standard of System of National Accounts (SNA). The major improvement of the 1970 input–output table is handling of sector classification. The industries in 1975 input–output table are expanded from 7 to 11, and the characteristic of 1975 input–output table is that endogenous sectors were divided into three groups including industry, producers of government services and producers of the private nonprofit services to household. By comparison, the manufacturing sector was substantially revised in 1980 and a new method of estimation of service sector is introduced in 1990. Furthermore, the 1995 input–output table expanded the service sectors and enhanced the basic materials for estimation. To reflect the changes of Japanese socioeconomic structure, some new sectors were embedded in the 2000 input–output table, such as "Reuse and recycling" and "Nursing care".

According to the annual report of International Input–Output Association in 2000, there are over 80 countries often constructing the input–output tables, including the Japan, Nederland, U.S. and other countries. As known, the main methods to construct the input–output tables include the survey-based method and non-survey-based method. Generally, each different method has outstanding characteristics. Survey-based method can guarantee the accuracy of the available data but at the expense of huge investment in terms of labor, time and financial resources. Most Chinese input–output tables are compiled with this method in the past. The counterpart, non-survey-based method, may spend much less. Currently certain countries construct the tables by using the hybrid method to reach the dual goals of both two previous methods. Theoretically, the hybrid method is somewhat a non-survey-based method.

2.2.1 Survey-Based Regional Input–Output Methods

In this section, firstly, the survey-based methods will be briefly introduced, such as cyclic census (complete survey), typical survey. Then a guideline with consecutive steps will be introduced to describe how to compile the input–output tables with the survey data.

2.2.1.1 Types of Survey-Based Methods

(1) Census (Complete survey)

A census is a complete enumeration of entire population as statistical units in a field of interest. For example, the population census canvases every household in a country to count the number of permanent residents and other characteristics, or a census of manufacturing may canvas all establishments engaging in manufacturing activities. The census of population (and households) is commonly carried out every ten years. The censuses of agriculture, fishery, forestry, construction, manufacturing, trade and other services are commonly carried out every five years. Similarly, a consumer income and expenditure survey is carried out every 5 years.

Data from the censuses serve as the base-year or benchmark data. A complete and up-to-date register of all statistical units in the field of inquiry is required. The advantages are that census provides the most reliable statistics if done professionally and with integrity, and the disadvantage is it is very costly to enumerate and to process data by means of a census. Timeliness is not high: data is available for use only many months, even years after it is collected. A census is normally carried every 5 or 10 years.

(2) Non-complete survey

Non-complete survey only focuses on some typical or key industries and sectors rather than all of the industries. There are three major types of this method including the sampling survey, key-point survey and typical survey.

Sampling survey is a method to randomly choose samples and estimate the total samples. It is used when there is no need or not able to adopt the complete survey. The sampling error may be controlled before implementation to ensure the data quality and every sample has the same probability to be selected. It is also used to evaluate the data quality collected by complete survey. This method can be divided into several types, such as simple sampling survey, stratified sampling survey, cluster sampling survey, systematic sampling survey and probability proportional to size sampling survey.

Key-point survey is a non-complete survey method to investigate the key industries, sectors or firms to get the information of the total sample. This method can be used in some areas in which some sectors overwhelmingly beat other sectors. This method can also be used to collect the data with quite low cost. This is much similar to the typical survey, which is to select some representative industries or sectors to project the situations and trends of these industries or sectors.

The differences of these three non-complete survey methods can be grouped into three parts. Firstly, the ways to choose the survey objectives are different. The sampling survey method should randomly choose the samples and everyone has the same probability to be chosen as a sample, no matter the potential survey objective is willing to or not. In point survey, choosing the samples is also objective. The indicators' value of the key industries should account for absolute proportion of the total value. However, in a typical survey, investigators have the own principles and criteria to choose the representative industries. Secondly, the purpose or the representativeness is different. As to the sampling survey method, there are certain rigorous and scientific calculation methods to project the total samples. Thus it can replace the complete survey method to some extent. The data collected by the point survey can only reflect the developing trends rather than the comprehensive information. The typical survey is frequently used to learn lessons from the sun-rise or sun-set industries. The data is hardly used to project the total samples due to lacking of robust scientific supports. Thirdly, according to the characteristics of survey objectives, specific survey method should be selected to collect the data.

2.2.1.2 Compilation Steps with Survey-Based Method

Both methodologies are still developmental to some extent as survey data gradually replaces non-survey data. Internationally, researches have started on constructing regional input–output tables in several countries including the Netherlands, Denmark, Italy, Canada and Finland. The regional input–output tables are based on the supply and use framework and are not derived square input–output tables. The regional supply and use tables are developed simultaneously with the national level ones. Because the national level input–output tables set the control totals for the regional level tables. This survey-based method is followed the constructing method implemented in China Statistic 2007. The general constructing processes of inter-regional input–output tables can be summarized to several steps.

(1) Regional tables of supply at basic prices
Where survey data can be used in this part and with undoubtedly these data can be subject to some potential problems. Specifically, the survey data may only cover certain industries rather than all the commodities. Further, the sum of the survey data collected from different survey methods may not be equal to the calculated national one. Nevertheless, if a good quality survey dataset does not exist, these can be obtained from those corresponding national tables, from which allocating the total value to specific regions based on some indicator variables, such as employment. Thereafter, the RAS technique and linear programming can be used to minimize the gap between regional and national proportions.

(2) Regional intermediate use tables at purchasers' prices
This is done the same as for supply.

(3) Regional final use tables at purchasers' prices
This is done the same as for supply but only for some considerations. The household consumption is estimated by using the disposable income of households by region. Consecutive yearly survey of the household expenditure can be used to reduce sampling error. In addition, the central government and local government final consumption is available from publications on state expenditure by region.

(4) Regional margins tables
Before calculating these tables, it is needed to assume that the margin can be divided into several regions based on the shares of total use of a particular good. So the researchers can thus allocate the national level tables into regions.

(5) Regional intermediate use tables at basic prices
Subtract the table of margins for each region from the regional intermediate use tables at purchasers' prices.

(6) Regional final use tables at basic prices
Subtract the table of margins for each region from the regional final use tables at purchasers' prices.

(7) Sort data by commodity to derive commodity account imbalances

(8) Trade flows for each commodity
A comprehensive multi-region targeted survey approach is taken in this step. Goods producing industries are surveyed for their sales and service industries are

surveyed for their purchases by region, which can be split into commodities considering that what commodities are export commodities and how these are allocated. The data is used to derive export and import flows by region for each commodity, among which the foreign imports are obtained as a residual.

(9) Balance of each commodity account
Each commodity account is balanced by calculating foreign imports as a residual.

(10) Finalization of regional industry-by-industry input–output tables.

2.2.2 Non-Survey-Based Regional Input–Output Methods

Non-survey techniques can derive elements of a regional input–output table from other (usually national) tables by various modification techniques, which use hybrid approach to combine non-survey techniques with superior data that obtained from experts, surveys and other reliable sources. Internationally, researches have contracted on regional input–output tables in several countries by using the non-survey-based methods, in which use the national level symmetrical input–output tables to derive regional tables. Specifically, the high level tables are adjusted by using indicators to calculate a region's contribution to the total industry aggregates.

At present the widely used non-survey-based methods include the RAS approaches and hybrid methods. The RAS technique requires less information than that of survey-based input–output tables. It is often regarded as a partial-survey or a non-survey method, in which some kinds of superior information (from small, focused surveys, expert opinion, etc.) are incorporated into an additional non-survey procedure. On the other hand, the hybrid methods often embed the regional table estimation problem in a large multi-region system. The regional input–output tables based on the hybrid methods mainly include Generation of Regional Input–Output Tables (GRIT).

The non-survey-based method is followed the constructing method of the Generation of Regional Input–output Tables (GRIT) implemented by the steps in Fig. 2.1. The general constructing processes of regional input–output tables can be summarized as the following steps:

(1) Update of the national input–output tables
In order to construct the regional input–output table, the national table must be updated for volume and price changes, which requires the combination of data from several sources. The international trade data is often used to update international imports and exports. Similarly, primary input and final demand figures are aligned to figures released by National Statistics Bureau. When the table is updated, table quadrants are balanced using the RAS technique. The updated

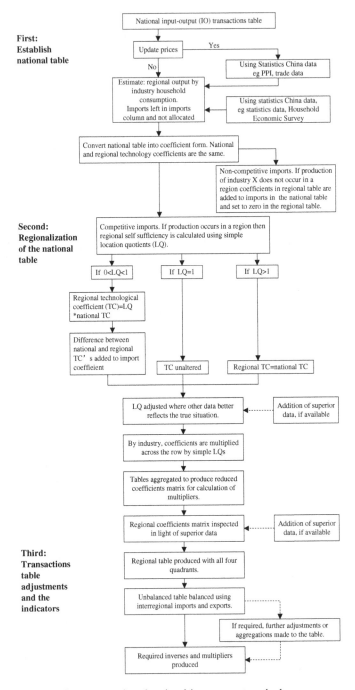

First:
Establish
national table

Second:
Regionalization
of the national
table

Third:
Transactions
table
adjustments
and the
indicators

Fig. 2.1 Summary of non-survey-based regional input–output methods

national table is converted into technical coefficient format with the assumption that the national and regional technologies applied in production are the same.

(2) Calculation of non-competitive imports

It is assumed that if the production in industry X does not occur in the region, then any inputs from industry X into industry Y are treated as regional imports. Thus, the technical coefficient in the relevant industry row is set to zero in the regional table, and the difference is added to the regional import coefficient.

(3) Calculation of competitive imports

This requires the estimation of self-supply in each regional industry, which is undertaken by using simple and cross industry location quotients. If the local supply cannot satisfy the demand in an industry then the imports are assumed to be required. The competitive imports are modeled by multiplying technical coefficients in the relevant industry row with the corresponding location quotient, and allocating the difference to the regional import coefficient. If local supply is able to satisfy local demand in an industry then the regional technical coefficient is set equal to its national equivalent.

(4) Calculation of industry aggregation

It is necessary to convert the regional technical coefficients into transaction values by using regional output estimates derived by multiplying national output figures by ratios of regional to national full-time equivalents (FTEs). The tables are aggregated to provide a reduced coefficients matrix for calculation. The coefficients are converted back to transactions values and sum the transactions. This is computationally easier than weighting the coefficients by output data and summing the weighted coefficients. Once expressed in transaction values industries may be aggregated as desired. Tables used for multiplier calculation are generally kept as disaggregated as possible to avoid aggregation bias from affecting multiplier estimates.

(5) Tables balancing

The tables are balanced using inter-regional exports based on a supply-demand pool approach, which is a commodity balancing approach commonly used in input–output table construction. Balancing is however not required if the table is being produced purely for the generation of multipliers.

(6) The insert of superior data and knowledge

The insert of superior data and knowledge can be undertaken at almost any point in the above process. If the developer believes that a mechanically produced LQ is not reflective of the degree of self-sufficiency in an industry, say, because of productivity differences, then adjustments could be made. Similarly, if survey data is available then this could be included.

Table 2.1 Theoretical GRFT transactions table (million CNY)

Industry	Industry A	Industry B	Industry C	Final demand	Total output
Industry A	25	20	15	40	100
Industry B	14	6	10	20	50
Industry C	20	12	43	25	100
Primary input	41	12	32	12	97
Total Output	100	50	100	97	347

(7) Calculation of regional transactions matrix and inverse matrices
The regional transactions matrix is developed using linear optimization and constrained using quality data (Table 2.1).

2.3 The Water and Land Resources Integrated Input–Output Table

The water resource issues are influenced by various factors in terms of climate change, land use and land cover change and socioeconomic development. To sustain the holistic natural and human health, numerous researches attempt to address these issues from different study areas (Deng 2011a, b). The simple conventional input–output tables ignore the interrelationship between the economic activities and the resources depletion, it is therefore necessary to build the resource integrated input–output tables for small scale researches to reveal the impacts of economic development on the natural resources such as water and land. Though a number of respectable researches have been conducted on the high level and international researches to meet the national demand of natural resource negotiations, there are few studies on the low levels, such as the regional level especially the county level, to support the inter-regional and intra-regional analysis among the basic prefectures.

The most concerned natural resources are the water and land resources around the world. In this research, both these two resources, which are regarded as the production factors in the input–output tables, are embedded in the integrated input–output tables at the county level.

The main descriptions to embed the water and land resources into the input–output tables can be divided into two parts. Firstly, the amount of water and land resources consumed in the key sectors should be estimated. Generally, to sustain development of a particular industry needs certain water resource (amount) and this industry will occupy some land resource (area). This is the base of the calculation and estimation of resource consumption of each sector. Secondly, the prices of both two resources are used to calculate their economic values, which are regarded as the economic input on the natural resources related to the water and land.

2.3.1 The Water Resource Integrated Input–Output Table

The water resource consumption in an integrated input–output table can be divided into three parts including the consumption of the primary industry, the secondary industry, and the tertiary industry.

As to the water consumption in the primary industry, it is necessary to find out real water of each sector, which means that the water used in a specific sector is from natural supply or anthropogenic activities. For example, the crop farming industry needs extrinsic investment to get water, while most water consumed in forestry industry is assumed that the water is directly obtained from precipitation and surface runoff.

Additionally, in order to construct a value input–output table, the water price has to be identified for estimating economic input of multiple sectors. For this purpose, the land use in crop farming industry is divided into two categories including irrigated land and non-irrigated land. Therefore, the irrigation coefficients can estimate the water consumption in different land use types. It is difficult to estimate water price in the crop farming industry. Thus, a new method is developed to calculate the water prices. Firstly, the difference of water consumption in both two land use types can be calculated. Then the difference of gross economic output between these two types of land uses can be obtained in order to calculate the economic output per unit water resource. This is regarded as the water price of crop farming industry, which can be applied to calculate the economic water input of the crop farming industry. When considering the water resource consumption in the forestry sector, animal husbandry sector and fishery, it is difficult and necessary to distinguish whether these sectors need "water resource" to industrial sustainable development. Since most forest ecosystem and aquatic ecosystem depend on natural water resource such as precipitation and surface runoff to a large extend to sustainable development without much human intervention, these sectors thus are regarded that there is no water resource consumption. It means that the water resource has no economic value (price = 0) though these sectors consumed certain water resource. This can explain that the ecological water consumption can be considered as the water resource input of these particular sectors that related to corresponding ecosystems, such as forest ecosystems, grassland ecosystems and the aquatic ecosystem. But there are no robust methods to estimate the economic value of natural water resource.

$$WI_{crop} = \frac{GO_{irrigated} - GO_{non-irrigated}}{WA_{irrigated} - WA_{non-irrigated}} \times WA \tag{2.10}$$

where WI is the economic value of water consumption of crop farming industry, the GO represents the gross output of different land use types (irrigated and non-irrigated), the WA is the amount of water consumption of different land use types.

With regard to the water consumption of the other industries, the non-survey based method is used to calculate the water resource consumption of each sector,

Table 2.2 Regional industrial land users' guide

Name	Region	Plot ratio	Fixed investment	Land yield	Land use indicators		
					Production scale	Land indicators	
					Large	Medium	Small
Industry A							
Industry B							

in which the total water resource is allocated to each sector by using the water use coefficients. Numerous researches have been conducted to estimate the industrial water use coefficients of each sector by using the input output analysis. These research data can be used as references to obtain the water consumption of each sector. Since most industries and sectors of the secondary and tertiary industry are situated in the urban area, it can be assumed that the water price of a particular prefecture can be used to estimate the economic value of water consumption of each sector.

$$WI_i = GO_i \times C_i \times P \qquad (2.11)$$

where C is the water use coefficients of each sector, P is the water price, and the *WI* and *GO* are the economic value of water consumption and gross output.

The urban and rural population is used to estimate the domestic water consumption. The existing statistic data associated with population is used to project the future population size which will be utilized to estimate the domestic water consumption. People live in different areas have different lifestyles in which the water use patterns are various. This is taken into consideration when estimate the water use per person in rural and urban areas.

2.3.2 The Land Resource Integrated Input–Output Table

In order to study the impact imposed on land resource of economic development, there is need to construct land resource integrated input–output table which includes the balance equations of the supply and use of the production sectors. Furthermore, this can provide scientific support to land management. Thus, it is necessary to develop a method to calculate the influence coefficient and sensitivity coefficient of land resources and carry out a quantity analysis on the change of land resources. In this part, the *User Guide of Industrial Land Use* written by the corresponding Guandong institutions is used to get the various coefficients of land use and management (Table 2.2).

2.3.2.1 Basic Regulations

The land resource integrated input–output table needs to satisfy the basic rules. It means that the land area shall not be greater than the corresponding scale of production land use indicator in the user's guide, and strength of investment shall not be less than regional guidelines index. The investment intensity is the fixed assets investment of per unit area within the scope of land. Furthermore, the plot ratio shall not be less than regional guidelines index. When calculating the plot ratio of the building which is higher than eight meters, it needs double calculate the construction areas. Otherwise, the coefficient of project construction shall not be less than 30 %. The construction coefficient is a proportion of all kinds of buildings and structures used in production and direct services to total land areas within the scope of land. Finally, the land output, which is the revenue of per unit land area within the scope of land, shall not be less than regional guidelines index.

2.3.2.2 Economic Value of the Land Resource of Each Sector

The land price in different regions is different. Thus, it needs to identify the areas with different situations. Then the total areas of industries can be collected from the regional statistical yearbooks and other public resources. Secondly, the plot ratio and fixed investment are used as the indicators to calculate the land use coefficients of each sector. Thirdly, the land areas of each industry can be calculated by multiplying the coefficients with the total areas used by all industries. Thereafter, the economic value of land resource can be estimated of industries. The equation is as follows:

$$\delta_i = \frac{Q_i}{\sum\limits_{i=1}^{n} Q_i} \tag{2.12}$$

where δ_i represents the coefficient of industry i in industrial gross output, Q_i is the land yield.

$$D_i = S\delta_i \tag{2.13}$$

where S is the total areas of all industries, D_i is the land area of industry i.

$$V_i = D_i P_i \tag{2.14}$$

where P_i is the land price, and V_i is the land value of industry i.

References

Deng, X. (2011a). *Modeling the dynamics and consequences of land system change.* Berlin: Springer, Berlin: GmbH & Co. KG.

Deng, X. (2011b). *Environmental computable general equilibrium model and its application.* Beijing: Science Press (in Chinese).

Harris, R.I., Liu, A. (1998). Input–output modelling of the urban and regional economy: the importance of external trade. *Regional Studies 32*(9), 851–862. doi: 10.1080/00343409850118004

Ihara, T. (1979). An economic analysis of interregional commodity flows. *Environment and Planning A, 11*(10), 1115–1128.

Miller, R. E., Blair, P. D. (2009). Input-output analysis: foundations and extensions. Cambridge University Press.

Norcliffe, G. B. (1983). Using location quotients to estimate the economic base and trade flows. *Regional Studies, 17*(3), 161–168. doi:10.1080/09595238300185161.

Chapter 3
Compilation of Regional Input–Output Table

3.1 Preparatory Work

In order to do a better job in an input–output survey, the National Bureau of Statistics, the National Development and Reform Commission and the Ministry of Finance jointly issued *The Notice of Conduct National Input–Output Survey Work*, which stressed that the input–output survey is the important foundation of compilation of national and regional input–output table. Input–output table is an important part of national economic accounting system. It is often the most powerful tool in carrying out policies for quantitative analysis, and be important to make managerial decision at macroeconomic level.

The compilation of input–output table relies on three approaches: the production approach, the income approach and the final expenditure approach. Each of those approaches requires a different set of data. The best practice is to combine all of them simultaneously in the framework of input and output tables. The main objective of that best practice is to avoid discrepancies in the three values of GDP volume obtained by applying three different methods separately. Thus, the compilation relies not only on data collected but also on aggregates, such as value-added and GDP, obtained as residuals through the national accounts compilation process. In addition, the balancing technique applied in balancing input and output tables would yield information concerning on the elements which statisticians do not have direct information or when it is too costly to collect information directly. For example, according to national accounts handbook of USA, grain production may be produced by numerous households but also by a few large corporations. The total output of grain is normally measured by the total crop area and estimated yield per acre. The total output of grain by corporations must be obtained by direct survey, but the total output of grain by households can be obtained as a residual. The total output of grain is then balanced with change in inventories, the intermediate use of grain in animal farming, a few manufacturing industries, and imports and exports of grain in order to obtain the total household consumption of grain. Thus, it is not necessary to survey households on their production and final

X. Deng et al., *Integrated River Basin Management*,
SpringerBriefs in Environmental Science,
DOI: 10.1007/978-3-662-43466-6_3, © The Author(s) 2014

consumption of grains. In regard of the above problem, scientific questionnaire and enough preparation are necessary.

The characteristics of input–output survey are strong comprehensive, wide range and difficult techniques. It needs reasonable and scientific division of labor organization. In order to ensure the progress of input–output survey done smoothly for getting the basic unit investigation data more accurate, the statistical departments at different levels have to be investigated under a well-prepared condition of filling in the questionnaire at the grass-roots level firstly.

1. Recognition of the significance of survey work

   The accounting of input and output includes three aspects, such as input–output survey, input–output table compilation and input–output analysis and application. Survey work is the basis preparation of input–output table. The quality of investigation data is related to the quality of the data of inputs and outputs, and it has a significant influence on macroscopic and microscopic qualitative and quantitative analysis of input–output table, economic policy, the macroeconomic regulation and control and enterprise management.

   It is easy to produce some misunderstanding to fill in the input–output table. For example, the investigation is just to meet the demand of the higher leadership authority, which did little to enterprise management. Owing to recognition of that it needs to work too much to fill in the input–output table, relevant entities and personnel need to seriously study the relevant documents and input–output survey system, especially the documents of National Bureau of Statistics and the National Development and Reform Commission and Ministry of Finance. Through the study, it can further promote input–output survey and prepare the input and output list that is the need of macroeconomic management and decision making. Additionally, the input–output survey is important to study the development of the national economy. By doing this, the state of enterprises and institutions can be identified in the input–output survey which is beneficial to promote enterprises improving internal accounting, comprehensive understanding the operating conditions of the units. Thus, it also facilitates strengthening management and improving the economic benefit by filling in input–output survey at the grass-roots level.

2. Main task of investigators

   The investigators are responsible for leading and organizing the units to provide input–output survey form at the grass-roots level. They should provide the basic input–output files, survey requirements and instructions of the personnel who participate in the input–output survey and learning. To ensure that the survey goes smoothly, the fill method should be developed and the conflict between the daily work and the survey work needs to be handled. In addition, it is necessary to set the related job schedule plan and clearly defined roles and responsibilities under the unified arrangement, collaboration and cooperate with each other. Thereafter, the data quality should be evaluated to reduce survey biases.

3. Reasonable arrangement of labor division

As input–output survey needs the statistics data in a wide range of industries such as production, technology, accounting, supply and marketing and so on, it is easy to duplicate or miss in the process of based information gathering and processing. Thus before filling in the input–output survey form at the grass-roots level, the investigators should study the resource of basic survey materials carefully and arrange the task of various departments reasonably according to the requirement of the input and output basic survey.

Generally speaking, a reasonable division of labor contains several parts. Firstly, comprehensive statistics departments are responsible for developing the guidance scheme of each unit, designing the statistics and calculation table, setting a unified index diameter and calculation method, providing the relevant statistical yearbook data, and making the report balance. Secondly, financial departments shall be responsible for providing the detailed information such as all kinds of costs, expenses and profit and loss data which can meet the demands of the requirement of input–output survey at the grass-roots level. Thirdly, material supply and marketing departments shall be responsible for providing and sorting the information about purchasing and using the materials for production and product sales. The last, construction department shall be responsible for providing and sorting the data of investment in fixed assets.

4. Staff training and set fill scheme

Staff training refers to a training activities that organized by the enterprises and institutions at the grass-roots level to the investigators and relevant participants who carry out the input–output survey and provide information. The experienced staff who participated in the province or city input–output survey is responsible for teaching and coaching. The content of training include the instructions and survey table of input–output table, the standard classification and code and related material of input–output table.

In order to accomplish the project of fill in the primary input and output questionnaire, a detailed fill scheme should be set. A clear schedule requirement can be proposed to make the working departments and specific tasks of the individual clear. The specific scheme should include five steps.

- The source and obtaining solution of data
- The department and individual that provide the detailed information
- The principle and calculation method of each index
- The summary procedure of literature and data review and the specific personnel
- According to the submission time requested by superior determine the progress of the unit that you responsible for.

3.2 Questionnaire Structure and Fill Method

The key questionnaire, in accordance with the investigation object is divided into 18 sets related to investment composition of fixed assets, industry, construction, wholesale and retail, road transport, water transport, air transport, accommodation, catering, software and information technology service, monetary and financial service, capital market service, insurance, illegal operation of real estate, human health, entertainment, other service and administrative institutions.

At the regional and county level, the major questionnaires are mostly related to investment composition of fixed assets, industries, construction, wholesale and retail, accommodation, catering, administrative institutions.

The prefecture level questionnaire mainly considers the investment in fixed assets, key industries, construction, wholesale and retail, accommodation, catering and administrative institutions. In this part, the industrial enterprise above designated size (sales revenue >20 million CNY) is taken as an example to introduce the design of regional industrial enterprise questionnaire, the data integration, revision.

1. Data source and fill method

    The production cost of industrial enterprises above designated size includes the industrial production and the manufacturing cost. The investigators only need to fill in the total amount of raw materials and intermediate products.

    The material sources of industrial enterprises above designated size include the purchase materials refers to raw materials, outsourced intermediate products and fuels. However, the production sales should not include the purchases sales. When fill in the questionnaire, the investigators should firstly list the names of the productions, the sales to domestic provinces and other foreign countries, and then summarize and verify the submitted data according to the national economy industry.

2. Questionnaire Structure

The questionnaire of industrial enterprises above designated size is taken as the sample as shown in Table 3.1.

3.3 Steps of Input–Output Table Compilation

The purpose of this section is to introduce the basic input–output system and construction method of the tables. This table considers a simple economy with a household sector and three industries producing three commodities. It is worth noting that the economy is purchasing a few imports, but has no exports.

Table 3.2 shows the money flows associated with the output of the three commodities. The agriculture and services industries are producing their characteristic outputs only. Manufacturing is producing services as well as manufactures.

Table 3.1 The key structure of questionnaire of industrial enterprises above designated size

Indexes	Code	Enterprise value	Small amount of products		
			Small products name	...	...
First	Second	1	Small products code	...	...
Total industrial value	1	–	–	–	–
Production cost	2	–	–	–	–
1. Direct materials consumption	3	–	–	–	–
Raw materials and supplies	4	–	–	–	–
Name Material classification		–	–	–	–
		–	–	–	–
Fuel and power	5	–	–	–	–
Bituminous coal and anthracite coal	6	–	–	–	–
Lignite coal	7	–	–	–	–
Other coal	8	–	–	–	–
Nature gas	9	–	–	–	–
Crude oil and oil products	10	–	–	–	–
Synthetic crude oil products	11	–	–	–	–
Coke and secondary product	12	–	–	–	–
Electric power and heat	13	–	–	–	–
Other	14	–	–	–	–
Packing product	15	–	–	–	–
Repair parts list	16	–	–	–	–
Other direct material	17	–	–	–	–
2. Direct labor	18	–	–	–	–
3. Other direct costs	19	–	–	–	–
Individual	20	–	–	–	–
Government	21	–	–	–	–
4. Manufacturing expenses	22	–	–	–	–
Managers pay	23	–	–	–	–
Managers of welfare funds	24	–	–	–	–
Depreciation expense	25	–	–	–	–
Repair charge	26	–	–	–	–
Commercial rent expense	27	–	–	–	–
Rental charge	28	–	–	–	–
Insurance expenses	29	–	–	–	–
Heating fees	30	–	–	–	–
Transport charge	31	–	–	–	–
Labor protection fees	32	–	–	–	–
Health care subsidies	33	–	–	–	–
Tool amortization	34	–	–	–	–
Design and drawing fees	35	–	–	–	–
Research expense	36	–	–	–	–

(continued)

Table 3.1 (continued)

Indexes	Code	Enterprise value	Small amount of products		
			Small products name	...	...
First	Second	1	Small products code	...	...
The test for inspection fee	37	–	–	–	–
Utility bills	38	–	–	–	–
Water bill	39	–	–	–	–
Supplies consumption	40	–	–	–	–
Travel expense	41	–	–	–	–
Office allowance	42	–	–	–	–
Service fee	43	–	–	–	–
Labor dispatch fee	44	–	–	–	–
Wages, social insurance premium	45	–	–	–	–
Labor management fee	46	–	–	–	–
Service fee	47	–	–	–	–
Postal and communication charges	48	–	–	–	–
Postal charges	49	–	–	–	–
Internet access fees	50	–	–	–	–
External processing fee	51	–	–	–	–
Social insurance charges	52	–	–	–	–
Housing fund and allowance	53	–	–	–	–
Other manufacturing expense	54	–	–	–	–
Individual	55	–	–	–	–
Government	56	–	–	–	–
Supplementary indicators		–	–	–	–
Sales expenses	57	–	–	–	–
Management Fees	58	–	–	–	–
Financial expenses	59	–	–	–	–
Average number of persons(person)	60	–	–	–	–
Accrued Wages	61	–	–	–	–

Table 3.3 shows the money flows associated with the use of commodities and the cost structure of the industries. The total use of each commodity in Table 3.3 must equal the total supply of each commodity in Table 3.2. Gross output of each industry in Table 3.3 equals the sum of its use of commodities plus primary inputs. Primary inputs include operating surplus, usually calculated as a residual.

Before deriving an inter-industry transactions table from the supply and use tables, the secondary production of services by the manufacturing industry should be addressed. Homogeneity of production is a key requirement for the analysis of

Table 3.2 The supply of commodity (million CNY)

Commodity	Industry			Imports	Total supply
	Agriculture	Manufacturing	Services		
Agriculture products	20	0	0	0	20
Manufactures	0	16	0	4	20
Services	0	4	30	6	40
Gross output	20	20	30	10	–

Table 3.3 The use of commodity (million CNY)

Commodity	Industry			Households	Total use
	Agriculture	Manufacturing	Services		
Agriculture products	4	10	2	4	20
Manufactures	4	2	10	4	20
Services	6	4	6	24	40
Primary inputs	6	4	12	–	–
Gross output	20	20	30	–	–

money flows between industries, and it is also one of the assumptions underlying stable input–output coefficients. Along with a proportion of the manufacturing industry's inputs, the 4 million CNY productions of services by the manufacturing industry should be moved to the services industry.

It is possible that extreme assumptions in shifting inputs from one industry to another. The industry technology assumption assumes commodities produced by an industry have the same input structure, or, commodities will have different input structures depending on the produced ways of industry. An alternative assumption for removing secondary production is the commodity technology assumption. It assumes that a commodity has the same input structure in whichever industry it is produced. The commodity and industry technology assumptions are regarded as two extremes of a range of assumptions about commodity production.

A hypothetical example is made that applying the industry technology assumption to manufacturing, that means 4/20ths or 20 % of all inputs of manufacturing get shifted to the services industry. The amended supply and use tables are shown in Tables 3.4 and 3.5.

The amended supply and use tables combine to produce the inter-industry transactions table which shows the buys (and sells) records between one industry and every other industry. The flows in the real world knowledge between particular industries, commodities can be refined to match those flows. Indeed such knowledge does not exist, proportions need to be used. Proportions require that the use of commodities in Table 3.6 needs to be weighed against the industries and imports that supply them. For example, the services industry produces 34/40ths or 85 % of total services, while imports contribute 6/40ths or 15 % of total services. See Table 3.6 for the derived proportions from Table 3.5.

Table 3.4 The supply of commodity adjusted for secondary production (million CNY)

Commodity	Industry			Imports	Total supply
	Agriculture	Manufacturing	Services		
Agriculture products	20	0	0	0	20
Manufactures	0	16	0	4	20
Services	0	0	34	6	40
Gross output	20	16	34	10	–

Table 3.5 The use of commodity adjusted for secondary production (million CNY)

Commodity	Industry			Households	Total use
	Agriculture	Manufacturing	Services		
Agriculture products	4	8	4	4	20
Manufactures	4	1.6	10.4	4	20
Services	6	3.2	6.8	24	40
Primary inputs	6	3.2	12.8	–	–
Gross output	20	16	34	–	–

Table 3.6 The supply table adjusted for secondary production, and showing market share of commodities (million CNY)

Commodity	Industry			Imports	Total supply
	Agriculture	Manufacturing	Services		
Agriculture products	1.0	0.0	0.0	0.0	1.0
Manufactures	0.0	0.8	0.0	0.2	1.0
Services	0.0	0.0	0.85	0.15	1.0

As shown in Table 3.5, the agriculture industry purchases 6 million CNY of services in total. The services industry provides 0.85 (from Table 3.6) or 5.1 million CNY of these, while imports contribute 0.15 or 0.9 million CNY. Table 3.6 also shows agriculture purchasing 5.1 million CNY of commodities from services. Agriculture also purchases 4 million CNY of manufactures. It purchases 0.8 or 3.2 million CNY from manufacturing and 0.2 or 0.8 million CNY are imported. There are no agricultural products imported and agriculture purchases its 4 million CNY of this commodity from itself.

There is a direct allocation of imports to industries and imports are no longer identified by particular commodity. The agricultural industry imports commodities worth 0.85 million CNY.

As shown in Table 3.7, all the inter-industry flows using proportionality assumptions. Manufacturing, for example, buys 8 million CNY of commodities from agriculture, 1.28 million CNY from manufacturing, 2.72 million CNY from services, imports 0.80 million CNY and adds value through its primary inputs of 3.20 million CNY.

Table 3.7 The inter-industry transactions table (million CNY)

Industry	Industry			Households	Total use
	Agriculture	Manufacturing	Services		
Agriculture products	4.00	8.00	4.00	4.00	20.00
Manufactures	3.20	1.28	8.32	3.20	16.00
Services	5.10	2.72	5.78	20.40	34.00
Imports	1.70	0.80	3.10	4.40	10.00
Primary inputs	6.00	3.20	12.80	–	–
Gross output	20.00	16.00	34.00	–	–

Table 3.8 Technology coefficients for industries table

Industry	Industry		
	Agriculture	Manufacturing	Services
Agriculture products	0.2000	0.5000	0.1176
Manufactures	0.1600	0.0800	0.2447
Services	0.2550	0.1700	0.1700
Imports	0.0850	0.0500	0.0912
Primary inputs	0.3000	0.2000	0.3765
Gross output	1.0000	1.0000	1.0000

Because the industries have been purified by removing secondary production, Table 3.7 can also regard as a commodity by commodity table. Industries and commodities are one and the same thing.

Table 3.7 can be converted to a table of technology coefficients for industries, which shows the proportion of inputs from industries and primary inputs to gross output. Agriculture, for example, produces its output with 4.00/20.00 or 20 % inputs from agriculture, 3.20/20.00 or 16 % inputs from manufacturing. Table 3.8 is a table of technology coefficients.

Suppose we want to know the effects of an increase in agriculture production on the other domestic industries. The increase in agriculture production is a result of an increase in spending on agricultural products by households. We need to consider the direct and indirect effects of the increased purchases of agricultural products by households. For example, 1 million CNY extra spending by households on agriculture will require a direct increase in agricultural output of 1 million CNY, but there are indirect requirements too. For agriculture to increase production it needs more inputs from manufacturing and services. These in turn will need some extra agricultural production, so the extra agricultural output required to boost household spending on agricultural products by 1 million CNY is likely to be considerably higher than 1 million CNY. The proportions in the shaded part of Table 3.9 are used to calculate the total requirements.

The system can be expressed as three equations:

Table 3.9 Industry by industry total requirements (direct and indirect) per unit of final demand

Industry	Industry		
	Agriculture	Manufacturing	Services
Agriculture products	1.6073	0.9683	0.5132
Manufactures	0.4345	1.4114	0.4777
Services	0.5828	0.5866	1.4603

$$0.2000X_1 + 0.5000X_2 + 0.1176X_3 + Fa = X_1$$
$$0.1600X_1 + 0.0800X_2 + 0.2447X_3 + Fm = X_2 \qquad (3.1)$$
$$0.2550X_1 + 0.1700X_2 + 0.1700X_3 + Fs = X_3$$

where X_1 is agricultural output, X_2 is manufacturing output, X_3 is service output and **Fa** is household purchases of agricultural output, **Fm** is household purchases of manufactures and **Fs** is household purchases of services.

The system can be written in matrix form

$$AX + F = X \qquad (3.2)$$

where **A** is the technology matrix. Rewriting with **X** as the subject

$$X = 1/(I - A)F \qquad (3.3)$$

where **I** is the identity matrix. The matrix **1/ (I-A)** is sometimes called the Leontief inverse.

The system of simultaneous equations can be solved (must be a square matrix) for any set of final demand values and gives us necessary industry outputs, which will satisfy any increase in final consumption.

Specifically, the Leontief inverse yields a set of new coefficients that show the increase in an input required to increase the final demand of an output by one unit. The direct and indirect requirements are shown in the Leontief inverse in Table 3.9.

As shown in Table 3.9, extra output of 0.9683 million CNY will be required from agriculture for manufacturing to increase its output by 1 million CNY for household consumption. In order to satisfy the indirect requirements of the other industries supplying manufacturing as well as the 1 million CNY direct demand by households, manufacturing actually has to increase its own production by 0.4114 million CNY. The diagonal values of the direct and indirect requirements table must, therefore, be greater than or equal to one.

The Leontief inverse is a powerful mathematical tool for unravelling the economy's complex interrelationships. It is the key to calculations which can reveal the original producers and ultimate consumers in an economy. Other useful tables are obtained using the Leontief inverse. The main ones are the 'cumulated primary input coefficients' and 'ultimate disposition of output' tables. These are derived by applying the Leontief inverse to the primary inputs and final demands quadrants, respectively, of the inter-industry transactions table.

3.4 Data Collection Methods

There are several important parts regarding data for regional input–output tables constructing. The data confidentiality will be a problem when publish data at regional levels. For example, if there are a small number of business units of a particular industry in the targeted region, the record data may be aggregated with another industry to a higher level within the unique industry classification, or it may not be published.

The data collection method must be designed at a regional level rather than a national level. It is necessary to design a large enough sample size to mitigate the sample errors. Of course, non-sampling error, which includes processing errors and non-response biases, may be presented in both sampling survey and administrative data. Non-sampling error is difficult to measure because it is unknown when or where it occurs due to an incomprehensive survey design and, therefore, the influence on data quality is unpredictable. In addition, it needs to find a way to identify the region to which the data relates, which means that all data must include some kind of regional identifier, for example, a business that is coded to a particular region.

Surveys can be conducted on different cycles, such as quarterly basis, annual, multiple-yearly cycles. Some surveys are conducted to get the latest data regularly to update existing information, while others, such as the census of population and dwellings are conducted every 5 years due to a statutory requirement. Theoretically, all data should be obtained from relevant resources simultaneously to construct the regional input–output tables. Certainly, data updated from several survey periods is possible if the economy is not undergoing significant changes, especially the prices, which can impact the productivity ratios. Data sources will probably include administrative data. If data is not available at regional levels then modeled data may be utilized.

Additionally, though the national level input–output tables only consider the international imports and exports, the regional input–output tables need to take the international and inter-regional imports and exports into consideration. The availability of inter-regional trade flows assists the production of balanced regional input–output tables and expenditure-based regional GDP (economic value added). Also, inter-regional trade flows highlight regional economic interdependencies and provide information for the measurement of the effects of regional and national economic policies. International trade flows need to include the region of origin for exports and the region of destination for imports. Without this information, regional input–output tables could not be balanced across regions, and important

Table 3.10 Information needs and data source evaluation for regional input–output table

Information need	Industry production accounts: Industries (excluding below), Farming-related industries, Central government
	Local government Primary inputs: Operating surplus, Compensation of employees, Taxes on products and imports, Consumption of fixed capital, Imports, Import duly
	Final demand: Final consumption expenditure: Households, NPISHs, Central government services, Local government services, Changes in inventories
	Gross Fixed Capital Formation: Exports, Inter-regional trade
Main potential existing data sources	Annual Enterprise Survey(AES), IR10s (tax data)
	Agriculture Production Survey
	Central Government Enterprise Survey, Crown Financial Information System
	Local Authority Census
	Derived from production accounts as a residual AES
	AES, and trade data.
	Household Economic Survey, Retail Trade Survey, and other surveys
	Annual organization accounts and other surveys
	Crown Financial Information System
	Local Authority Survey
	AES, Central Government Enterprise Survey, Local Authority Census, Wholesale and Retail Trade Surveys, Agriculture Production Survey, and other surveys
	Central Government Enterprise Survey, Local Authority Census, Quarterly Building Activity Survey, Agriculture Production Survey, and other surveys
	Trade data.
Possible change for regional data	Additional survey questions to collect data at geographic unit level-investigate IR10 data for level of geographic unit
	Some issues to address, but no suggested changes
	Allocation of central government data across regions by surveying distribution of government expenditure
	No suggested changes
	Not applicable
	Use the Linked Employee Employer Database and/or Quarterly Employment Survey data
	Imports comments
	AES comments
	Addition of questions regarding destination of imports
	Use the Regional Household Expenditure Database
	See Central Government comments
	No suggested changes
	See AES, Central Government comments—increase the sample size of the Wholesale Trade Survey and collect origin and destination data of sales and purchase
	See Central Government comment—increase the sample size of the Quarterly Building Activity Survey
	Addition of questions regarding origin of exports
	Addition of questions to existing surveys or conduct a stand-alone survey

information regarding a region's import or export patterns could not be determined.

As to the data collection information, Table 3.10 summarizes the theoretical information for regional input–output tables and the data sources used for the national input–output tables that may provide information at a regional level. Generally, it contains the data from administrative records and collected by statistical methods.

3.4.1 Historical Records

Generally, the administrative records can be roughly divided into two categories based on the data users. Specifically, some records are prepared and submitted to higher authorities, such as government revenue and expenditure statistics, foreign trade statistics and money and banking statistics. Some reports on insurance companies collected by the insurance regulatory authority are also subjected to this type. Besides, the tax records and business accounts of publicly traded corporations can be also used by the higher authorities. On the other hand, the following records are prepared for internal uses by corporations, such as the business accounts of corporations that include the income statement, the change in the financial position or cash-flow statement, and the balance sheet. Also the market analyses conducted by producers associations can describe the internal uses of a particular corporation.

After gathering the recorded data, methods should be taken to evaluate the quality of data to ensure the high coverage and reliability. Since it takes much time to process the recorded data so that the timeliness of the data is quite low and the cost is considerable. In order to speed up data availability of administrative records, a sampling of tax records may be utilized. For example, the use of budgeted government revenues and expenditures may be corrected for implementing indicators. Further, revision of administrative records is needed when complete and audited data are available.

3.4.2 Statistical Survey

A wide range of statistics is collected by government for purposes other than national accounts by censuses and surveys. A census of all statistical units in a given population is carried out every 5 or 10 years. Every year or quarter, a sample is taken to estimate population data. The most important requirement for a sample survey to be reliable is that the register of statistical units is up to date.

The advantage of sampling survey is that it can estimate the total sample by using quite few random samples, which can reach the research goals with these methods and corresponding interval estimation. In contrast, the key-point survey and representative survey alike cannot do this. On the other hand, there is an ever-existing problem that it is no way to assure the representativeness in theoretical researches and practical studies, which make it stay at an embarrassing situation. This method is implemented at the huge cost of human and financial resource. After introducing the market economy system, the potential risk in statistic data is increasing due to the curbing investment into the statistic work. Hence, when the sampling survey being introduced, it is widely accepted and used in the practical studies including the regional sampling and catalogue sampling.

From the national level perspective, the survey site should be stretched to the county, street and even residential committee, which cost enormous resources. Thus, the regional sampling survey is not widely applied in the national level statistic survey. However, it is scientific and feasible to apply the regional sampling survey if a reasonable survey scheme and sampling method is designed in a specific region. Thus, it is necessary to implement a sampling survey for supplement to compiling the county-level input–output tables and collecting the relative socioeconomic, industrial and trade datasets accurately, which would facilitate deeper understanding to local economy structure than non-survey input–output compilation.

Given the considerable differences among individual business, family business, enterprises, industries and the scales, the regional sampling survey methods have their own characteristics of specific target.

1. Enterprise sampling survey
   This sample survey covers a broad range of firms who engage in production and service in the whole country or in a region. As the basic organization unit of production activities, enterprise is different from the individual business and family business. Therefore, compared with the surveys of individual and family level, the firm-level investigation has its clear characteristics. Firstly, the scales of firm are different. Some have hundreds and thousands of employees while the counterparts have only several people. Secondly, the investigation objects will change with the enterprise foundation, close down, break down and write off. Thirdly, the enterprises can be divided into various industries whose production factors, operation models and production output are different from each other due to the imparity of labor, technology, labor productivity and balance sheet ratios. Fourthly, the investigators can collect the registered capital, sales revenues and taxes from the Industrial and Commercial Bureau, which is help build the framework of sampling and estimate the total sample if take it as auxiliary variables. Fifthly, the data collected from the enterprise survey is an important data resource of System of National Accounts (SNA). Thus, the definitions, investigation coverage and classification should also meet the SNA's requirement. Therefore, the enterprise sampling survey is more complicated than the individual and family business survey. In order to reach a better result in the practical implementation, it is clearly necessary to take the above characteristics into consideration to design an appropriate sampling survey scheme.

Regional sampling survey is a relative simple method, and there is no need to select the standard values and worry about the representative mud. Taking the sales of urban commerce as an example, all you have to do is to divide the city map into same grid (100 cells), and label these cells with consecutive numbers. Then 10 cells will be random selected as the basic statistic unit to calculate the sales. The total sales thereafter can be estimated by times 10 with the sample results. If the wholesale and retail sales have to be estimated, the investigators can calculate the volume of each category and then time 10. As to the sample size (5, 10 or 20), it is determined by the investment of finance and time, and the calculation accuracy as well. In addition, the survey schedules differ from industry to industry. For instance, if the investigator has to enquire the industrial factories, it is necessary to identify the location, scale of the factory in different jurisdictions and at different levels (city or county level).

In order to meet the demand of investigation information at multiple levels, such as the national, provincial and prefectural level, it is necessary that choosing the samples at multiple levels for facilitating superior authorities to manage the investigation information and evaluate the estimation results of lower level institutions. Given the distribution of the firms, the county (prefecture) is taken as the basic unit of sampling survey. However, the sample sizes of different levels in a targeted county are different. Generally, the sample size at province level is less than that of the city level, which is also less than that of the county level. For the sake of money saving and material sharing principles, there should be same samples at multi-levels. It means that the samples at the city level should cover that of province level, and the county level samples cover that of the city level. Thus, there are two sampling methods. The first one is to choose the samples at the county level, among which the city level samples can be selected from the county level samples and then province level samples can also be identified. This is called the three-stage sampling method. The second one is to choose the province level samples at targeted counties, then add more factories to make the city level samples and then the county level samples can be selected if new factories are added to the city level samples. Considering the difficulties and complexities of the former to estimate the variance comparing with the one-phase sampling method with the same sample size, the latter is often chosen in researches.

As to the county level sampling survey, it is appropriate to use the stratified symmetric isometric sampling method. The county level industrial enterprises below designated size included in the sampling frame can be stratified based on the total output or total sales revenue of the last year. This method can enhance the representative of samples. Though these enterprises under designated size, the sales revenue differs from less than one million to over tens of millions. So the pure random sampling method may result in the sampling biases. If these industries are grouped into several levels (generally three levels), the samples selected by the stratified sampling method can represent the total sample structure.

The design of stratified symmetric isometric sampling method has advantages over several aspects. Besides satisfying the need of multiple levels, the authorities at each level spend the corresponding cost to conduct the investigation, which can reach the dual goals of responsibility distribution and material sharing and financial saving as well. Furthermore, the investigator can get information directly from the factories regardless of the intervention of intermediate linkage, which can ensure the data reliability. In addition, by comparing the estimation results at different levels, it can appraise and adjust the sampling method. It is flexible for the investigators at various levels to implement the survey to get the data at the required accuracy.

2. Sampling method of wholesale, retail and catering services
The wholesale, retail and catering services are important parts of national economy. Currently, both the complete survey and sampling survey are widely used to collect the statistic data, which means that the statistic-reporting system is applied to investigate the business unit above norm and the sampling survey is used to investigate the business unit under norm.
This type of sampling method has the following characteristics. The total sample is always changing, which cause the problems in the survey processes. It is difficult to determine and improve the sample frame due to the unstable amount of samples, which can also increase the uncertainties in the total sample estimation. Further, the sales revenue of different business units differs from each other from the perspective of geography and individual units. More important, there are some extremely sensitive economic indicators in the process of sampling survey, thus it is necessary to improve the survey design and survey skills to reduce the non-sampling biases.
Specifically, the wholesale and retail industry can be divided into two categories: market and non-market industries. The market industries with above billion CNY can be selected in the whole province using the Probability Proportionate to Size Sampling, while the counterparts can be selected by the regional sampling method. Accordingly, the latter one consists of four parts of sampling objects including the counties; business and commercial station; the commodity transaction market under billion CNY, communities and village committees; and booths under billion CNY and non-market business units. In fact, the non-market industries above billion CNY can be divided into two parts: sampling in the whole counties and sampling in the sampled counties.
As for the sampling within the whole counties, the business and commercial station is selected as the secondary unit who covers over one street, village and town. The reason why the business and commercial stations can be chosen as the unit is that they have the data of individual units with annual inspection, which is an ideal auxiliary variable. Certainly, as the alternative sampling method, the street, village and town can be choose as the secondary unit to conduct the sampling survey. In this method, only the population and other variable can be regarded as the auxiliary variables. Furthermore, the business and commercial units can also be grouped into market and non-market one. In

each group, the stratified sampling survey should also be used to gather the datasets.

(1) Sampling method of commodity transaction market above a billion CNY
Given the type of industries accounts for large proportion of the sales revenue and they have the catalogue, the sampling survey can be conducted in the whole province. The two-phase sampling method will be used: sample the market and then sample the business unit in the sampled market. Moreover, the amount of market above one billion CNY crossing various administrative ranges from hundreds to thousands, so as to have to set the smallest sample size for guaranteeing the sampling accuracy. When the total sample (N) is less than 10, all the market should be investigated, while N is larger than 10, the sample size will be calculated by the formulation $(3\sqrt{N})$. As to the market sampling, the systematic sampling method is applied to investigate the leased booths.

(2) Sampling method of counties
The methods used in this part are similar to industry survey method. The stratified sampling method is used in the first one, in which the simple random sampling method is applied to choose n counties, and the sample size is determined by the counties' number in each level. This one can be used to choose the province-level samples. If some areas of the province are supposed to be estimated, then the second survey method which can be called appended sampling method should be used. This method will adopt the sampling without replacement, which means that the additional added counties should not be the existing samples counties. The stratified sampling will be also used here. If the number of county in one level is less than 2, then all the counties in this level will be regarded as the samples. In contrast, if the number is larger than 2 and the number of sampled counties in the first method is less than 2, then the simple random sampling method will be used to choose the samples in the non-sampled ones.

(3) Sampling the business and commercial stations in the sampled counties
Sampling in the sampled counties is an important part and the sampling method is the same in each counties. The secondary sampling unit can choose the business and commercial station or street, village and town. The former choose the number of the individual units with annual inspection, while the latter can choose the population and other variable as the auxiliary ones. No matter which one will be used, the stratified systematic PPS sampling method should be utilized.

(4) Sampling the sampled business and commercial stations (street, village and town)
The next sampling unit of sample business and commercial station includes the market under one billion CNY, community and village committee. The stratified PPS sampling method is used in the market samples choosing in

which the leased booth is utilized as the auxiliary variable. The investigator should inquiry all if the number of markets is equal or less than 2, choose 2 if it is less than 5 and choose 3 if equal or larger than 6. In each market, the samples should be selected by the isometric sampling method.

As to the sampling of community and village committee, there is need to classify the non-market community and village committee at the first stage. In each level, the simple random sampling or the isometric sampling method can be used to choose 2 samples. Thereafter, the two-stage sampling will be applied to choose the samples in the sampled communities and village committee. The first stage is to count the total number of the business unit by using the cluster sampling, and then classify them based on the scale and business type. The second stage is the field survey of business units. It is necessary to investigate all if the number of business unit is less than 5. Or choose several units using the stratified random sampling method or stratified systematic sampling method.

Firstly, it is to estimate the total sampled counties, which include the market and non-market ones. The Thompson order formula is used to estimate the total number of market samples. As to the non-market samples, it is necessary to estimate the total number of sampled communities. Specifically, the total number is simply to add up all the samples if the sampled communities are less than 5. If it is larger than 5, the total number is to multiply the average number of stratified sampling in each level with the number of levels.

3.5 Identification of the Leading Industries

3.5.1 Concept of Leading Industry

The concept of leading industry in China is originated from *The Phase of Economic Growth* published by Rostow in 1960. Some other introductions are mainly originated from *Strategy of Economic Development* published by Hirschman in 1958. The current leading industry theory is the combination of both two concepts.

The Hirschman has investigated the leading industry selection standard. He put forward due to the scarce resources and lack of entrepreneur and infeasibility of the balanced growth. With the limited capital, the government should increase the imbalance of supply and demand and pay important attention to the investment of major industry and takes the initiative to develop the correlative industry to lead to the development of whole industry.

In the 1950s, Japanese industry economist has put forward the dynamic comparative cost theory in his published book *Theory of Industrial Structure* for the first time. The author thought that we shouldn't statically determine whether the product should be produced from the comparative cost of the product. Furthermore, to choose regional leading industry should compare its cost to make the right judgment from a dynamic perspective and based on that to determine whether it should be chosen as the regional leading industry.

3.5.2 Identification Methods

Leading industries play an important role in regional economy. Its development directly affects the economic level and regional industrial structure. In general, the leading industries in developed regions have a good future and will impose a strong linkage effects on other industries, while the counterparts are mostly malnourished in poverty-stricken area. However, concepts of the regional leading industries are variously given by economists, which mainly include the three ones.

1. Industry Life Cycle
   The division based on the different stages of industry life cycle. Four stages includes the input stage, growth stage, mature period and degenerating stage, and the leading industry regard as the growth-stage industry which can be characterized by high growth rate and high speed of development because it plays a key role in changes of the whole industrial structure. The reason why leading industry can break the original relative balanced industrial structure is that it creates and meets new social needs. In general, leading industry can often represent the new market demand, new direction of industrial structure transformation and a new development level of modern science and technology industrialization.

2. Industry Development Ordering
   According to the industry development sequence, the industries can be divided into basic industry, leading industry, high-tech industry. Leading industry is referred to the industry that it can widely affect the structure of other industries directly or indirectly and drive economic growth in some stages. Leading industry is an industrial cluster consists of several industrial sectors, and they can absorb advanced technology and scientific and technological innovation achievement, meet the market demand of rapid growth. Therefore, it can acquire higher productivity.

3. Linkage Effects
   The leading industries are also identified according to the difference in the industrial position and role of national economy. The leading industry is defined as the industry or industrial clusters that it can depend on scientific and technological progress or innovation to acquire new production function and it can also drive other related industries develop rapidly through rapider development out of proportion. These leading industries or industrial clusters possess at least three characters at the same time. First, it can depend on scientific and technological progress or innovation to acquire new production function. Second, it is able to form continuous high-speed growth. Third, it possesses strong

diffusing effect, namely, it plays a decisive influence role in driving the developing of other industries even all industries.

The widely accepted and used method in this theory is to estimate the linkage effects based on the input–output table by some indicators, such as the influence coefficient, sensitivity coefficient.

(1) Influence coefficient

The indicator is to estimate the impacts of a particular industry development on other industries, which is also known as the index of the power of dispersion. It reveals the input requirement of other industries if the final use in a particular industry increased by one unit. It is used to estimate the backward linkage effects. If RB_j is the influence coefficient, the computational formula is

$$RB_i = \frac{\sum\limits_{i=1}^{n} b_{ij}}{\frac{1}{n}\sum\limits_{i=1}^{n}\sum\limits_{j=1}^{n} b_{ij}} \quad (i,j = 1,2,\ldots,n) \tag{3.4}$$

where b_{ij} is the element of Leontief inverse matrix $(\mathbf{I} - \mathbf{A}) - \mathbf{I}$ in ith row and the jth column, $\mathbf{A}$ is the direct consumption coefficient matrix.

(2) Sensitivity coefficient

Sensitivity coefficient is an indicator to estimate the sensitivity or response of one particular industry if the final use of all the other industries increases one unit, which is also linked with the forward linkage effect. This is called the index of the sensitivity of dispersion, which can be calculated

$$RF_j = \frac{\sum\limits_{j=1}^{n} b_{ij}}{\frac{1}{n}\sum\limits_{i=1}^{n}\sum\limits_{j=1}^{n} b_{ij}} \quad (i = 1,2,\ldots,n) \tag{3.5}$$

where b_{-ij} is the element of Leontief inverse matrix $(\mathbf{I} - \mathbf{A}) - \mathbf{I}$ in ith row and the jth column, $\mathbf{A}$ is the direct consumption coefficient matrix.

(3) Technical coefficient

In input–output analysis, the technical coefficient is to identify the percentage of the total inputs of a sector required to be purchased from another sector. It also represents the direct backward linkages of an industry to other industries.

$$T = (t_{ij})_{1*n} \tag{3.6}$$

where T is technical coefficient matrix, $t_j = (V_j + M_j)/X_j$, and t_j is growth rate of added value.

(4) Economic profit coefficients

There are many indexes reflecting economic profits. We select some indexes based on the input–output table.

$$\omega = added\ value/total\ input \tag{3.7}$$

(5) Structure of labor input coefficient

$$L_j = V_j/X_j \tag{3.8}$$

where L_j is the percentage of the labor input in the jth account on total input.

(6) Industrial expansion coefficient

We select industrial added value to reflect the capacity of industrial expansion.

$$Industrial\ expansion\ coefficient = \frac{industrial\ added\ value}{total\ industrial\ added\ value} \tag{3.9}$$

Chapter 4
The Integrated CGE Model Construction

4.1 Model Framework and Function

CGE models are simplified representations of entire economies. One approach to constructing a CGE model is through the notion of the circular flow of an economy. Figure 4.1 presents the core of the conceptualized circular flow in a CGE model, adapted from Ghadimi (2007). First, we start with the producers. A CGE model contains multiple producing sectors such as the agricultural sector, the manufacturing sector, the trade sector, the services sector and the utilities sector. The number of sectors (and model complexity) could vary from only two sectors to hundreds sectors, depending on the level of aggregation of industry activity needed for a particular policy analysis. An integrated CGE model often includes a water utility sector that captures, stores, treats and delivers water for its customers. Each industry sector is represented in the model in aggregate over all firms and therefore with a specific production function.

Producers in a CGE model purchase inputs to produce commodities to sell in the product market. For example, the agricultural sector purchases fertilizer, seed, tractors, gasoline and so forth from the product market. These are called inter-industry purchases. Producers also purchase the services of factors of production. The agricultural sector, for example, purchases labor, capital and land from their owners. In the case of the integrated CGEs, water may also be considered as a production factor. An electric utility may own water rights that permit the utility access to a given percentage of water available in a given year and watershed. The electric utility pays a "rent" for the right to use this water in the same way that an agricultural producer might pay rent for the right to use land.

In the integrated CGE model, the owners of the factors of production are called households. Households may consist of a single representative household or, if different income levels, locations, ethnicities, or other characteristics are of interest to the modeler, more than one representative household. All factor income accrues to households as the ultimate owners of the factors. To complete the circle, households spend the income that they receive for the use of the factors they own in the product market. In the integrated CGE model, one of the commodities purchased in the product market may be water from the water utility sector.

X. Deng et al., *Integrated River Basin Management*,
SpringerBriefs in Environmental Science,
DOI: 10.1007/978-3-662-43466-6_4, © The Author(s) 2014

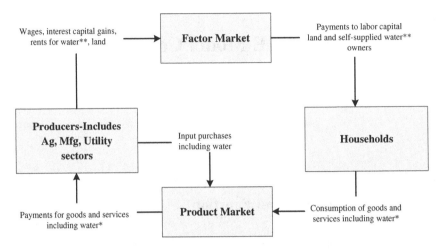

Fig. 4.1 Circular flow of income for the integrated CGE model. *Note* * Water "produced" by the water utility sector, ** Water owned as a factor of production, i.e. water supplied through ownership of water rights

While the above represents a basic description of the core of a CGE model, CGE models typically also contain representations of a government sector, investment and savings, and trade. The government sector is important to help model the lack of a market in the water factor market or the water commodity market. Governments collect taxes, consume commodities, and redistribute some taxes. In the case of some integrated CGE models, water 'prices' are specified as taxes or fees, as opposed to market clearing prices that are determined endogenously through the model, which are redistributed back to households. Investment and savings specifications become important for dynamic CGE models in order to connect savings and investment in the initial time-period with capital formation. This can be especially important for dynamic integrated CGEs that consider policy questions about water supply infrastructure over time. Specification of trade flows with other regions are a standard part of CGE models and may also be important for modeling trade liberalization in conjunction with changes in the institutional structure for water rights, as well as in multi-region.

4.2 Equations Included in the Integrated CGE Model

4.2.1 Price Equations

The price system of the model is rich, primarily because of the assumed quality differences among commodities of different origins and destinations (exports, imports, and domestic outputs used domestically). The price block consists of equations in which endogenous model prices are linked to other prices (endogenous or exogenous) and to non-price model variables.

$$PM_{row,c} = (1 + tm_{row,c}) \times pwm_{row,c} \times EXR \qquad (4.1)$$

where
$c \in C$ set of commodities
$row \in ROW$ set of foreign trading partner countries
PM_{rowc} domestic import prices with margins and tariffs in LCU (local-currency units)
tm_{rowc} tariff rate on imports
pwm_{rowc} import prices in FCU (foreign-currency units) (CIF or CFR.)
EXR foreign exchange rate (LCU per FCU).

The import price in LCU (local-currency units) is the price paid by domestic users for imported commodities (exclusive of the sales tax). Equation (4.1) states that it is a transformation of the world price of these imports, considering the exchange rate and import tariffs.

$$PE_{c,row} = (1 - te_{c,row}) \times pwe_{c,row} \times EXR \qquad (4.2)$$

where
$PE_{c,row}$ domestic export prices with margins and subsidies in LCU (local-currency units)
$te_{c,row}$ tax rate on exports
$pwe_{c,row}$ export prices in FCU (fob or exw).

The export price in LCU is the price received by domestic producers when they sell their output in export markets. This equation is similar in structure to the import price definition.

$$PQ_c \times QQ_c = \left[PD_c \times QD_c + \sum_{row} \left(PM_{c,row} \times QM_{c,row} \right) \right] \times (1 + tq_c) \qquad (4.3)$$

where
PQ_c composite commodity price
QQ_c armington composite commodity
PD_c price for commodity produced and sold domestically
QD_c commodity produced and sold domestically
$PM_{c,row}$ price for import commodity
$QM_{c,row}$ quantity of import commodity
tq_c rate of sales tax.

Aggregated absorption is total domestic spending on a commodity at domestic prices corresponding to its final demand. Equation (4.3) defines it exclusive of the sales tax. Absorption is expressed as the sum of spending on domestic output and imports at the demand prices, PD and PM. The prices PD and PM include the cost of trade inputs but exclude the commodity sales tax.

$$PX_c \times QX_c = PD_c \times QD_c + \sum_{row} \left(PE_{row,c} \times QE_{row,c} \right) \qquad (4.4)$$

where
PX_c aggregated producer price
QX_c commodity output
$PE_{row,c}$ price for export commodity
$QE_{row,c}$ quantity of export commodity.

For each domestically produced commodity, the marketed output value at aggregated producer prices (PX) is stated as the sum of the values of domestic sales and exports.

$$PA_a = \sum_c \vartheta_{a,c} PXAC_{a,c} \qquad (4.5)$$

where
$a \in A$ set of activities
PA_a production activity price (unit gross revenue)
$PXAC_{a,c}$ price from activity to commodity
$\vartheta_{a,c}$ yield of output c per unit of activity.

The gross revenue per activity unit, the activity price, is the return from selling the output or outputs of the activity, defined as yields per activity unit multiplied by activity-specific commodity prices, summed over all commodities. This allows for the fact that activities may produce multiple commodities.

$$cpi = \sum_c cwts_c PQ_c \qquad (4.6)$$

where
cpi consumer price index
$cwts_c$ Initial share of investment on commodity or weight of commodity c in the CPI.

Note that the notational principles make it possible to distinguish between variables (upper-case Latin letters) and parameters (lower-case Latin letters). This means that the exchange rate and the domestic import price are flexible, while the tariff rate and the world import price are fixed. The fixedness of the world import price stems from the "small-country" assumption. That is, for all its imports, the assumed share of world trade for the modeled country is so small that it faces an infinitely elastic supply curve at the prevailing world price.

4.2.2 Production-Related Equations Defined at the National Level

Production is carried out by activities that are assumed to maximize profits subject to their technology, taking prices (for their outputs, intermediate inputs, and factors) as given. In other words, it acts in a perfectly competitive setting. The CGE model includes the first-order conditions for profit-maximization by producers. As noted in the preceding section, two alternative specifications are permitted at the top level of the technology nest: the activity level is either a CES or a Leontief function of the quantities of value-added and aggregate intermediate input use.

1. Sectorial production function

$$X_i = A_i L_i^{\delta_i} K_i^{\kappa_i} N_i^{\eta_i} W_i^{\omega_i} \tag{4.7}$$

where

A_i shift parameter in sector i
L_i production factor of labor in sector i
K_i production factor of capital in sector i
N_i production factor of land use in sector i
W_i production factor of water resource consumption in sector i
δ_i elasticity of labor input for output in sector i
κ_i elasticity of capital input for output in sector i
η_i elasticity of land use for output in sector i
ω_i elasticity of water resource consumption for output in sector i

2. CES top-level production function

$$QX^1 = A_1 \left(\alpha_1 QF_{cap}^{-\rho_1} + (1 - \alpha_1) QF_{lab}^{-\rho_1} \right)^{\frac{1}{-\rho_1}} \tag{4.8}$$

where

QX^1 the first level composite quantity
A_1 shift parameter in the CES function
QF_{cap} quantity of capital input
QF_{lab} quantity of labor input
α share parameter in the CES function

$$QX^2 = \frac{A_2}{1 - \sum lc}\left(\alpha_2 QF_{lnd}^{-\rho_2} + (1 - \alpha_2)QX_1^{-\rho_2}\right)^{\frac{1}{-\rho_2}} \tag{4.9}$$

where
QX^2 the second level composite quantity
A_2 shift parameter in the CES function
QF_{lnd} quantity of land use

$$QX^3 = \frac{A_3}{1 - \sum lc}\left(\alpha_3 QF_{wtr}^{-\rho_3} + (1 - \alpha_3)QX_2^{-\rho_3}\right)^{\frac{1}{-\rho_3}} \tag{4.10}$$

where
QX^3 the third level composite quantity
A_3 shift parameter in the CES function
QF_{wtr} quantity of water resource consumption

3. CES aggregated Leontief top-level production function

$$QX_a = ad_a^x\left(\delta_a^x QVA_a^{-\rho_a^x} + \left(1 - \delta_a^x\right)QINT_a^{-\rho_a^x}\right)^{-\frac{1}{\rho_a^x}} \tag{4.11}$$

where the value added production function (QVA_a) is

$$QVA_a = iva_a QX_a \quad \text{and} \quad QINT_\alpha = inta_\alpha QX_\alpha \tag{4.12}$$

the intermediate demand $(QINT_c)$ is

$$QINT_c = \sum_{a \notin APIR}\left(io_{c,a} \times QA_a\right) + \sum_{a \in APIR}\sum_{i,r}\left(fio_{i,r,c,a} \times FQA_{i,r,a}\right)$$
$$+ \sum_{a \in APIR}\sum_{i,r} QINT\_G_{i,r,c,a} \tag{4.13}$$

where
$i \in I$ water administrative districts (basin)
$r \in R$ water administrative regions
$APIR \in A$ activities inside water districts (basin)
$QINT_c$ aggregated intermediate demand
$FQA_{i,r,a}$ production activity quantity at the basin level
$QINT\_G_{i,r,c,a}$ energy demand for intermediate input in ground water production
                      function at the basin level
io_{ca} input–output coefficient
$fio_{i,r,c,a}$ input–output coefficient at the perimeter level

4. Aggregated Leontief second-level value added and intermediate input functions

$$QVA_a = ad_a^{va} \prod_{f \in F} FD_{f,a}^{\alpha_{f,a}^{va}} \qquad (4.14)$$

$$WF_f = \frac{\alpha_{f,a}^{va} PV_a QVA_a}{FD_{f,a}} \qquad (4.15)$$

$$QVA_a = ad_a^{va} \left[\sum_f \delta_{f,a}^{va} FD_{f,a}^{-\rho_a^{va}} \right]^{-\frac{1}{\rho_a^{va}}} \qquad (4.16)$$

$$WF_f = PV_a QVA_a ad_a^{va} \left[\sum_f \delta_{f,a}^{va} FD_{f,a}^{-\rho_a^{va}} \right]^{-1} \left[\delta_{f,a}^{va} FD_{f,a}^{-\rho_a^{va}-1} \right] \qquad (4.17)$$

where
WF_f factor price, including water price and land rent
$FD_{f,a}$ final demand of i sector

5. Relationship between value-added and activity prices

$$(1 - ta_a)PA_a = PVA_a + \sum_c \left[io_{c,a} \times PQ_c \right] \qquad (4.18)$$

where
PVA_a value-added price, net intermediate input cost
PA_a intermediate input price, production activity price (unit gross revenue)
PQ_c intermediate commodity price
ta_a tax rate on production activity.

4.2.3 Production-Related Equations Defined at the Subnational Level

1. Factor demand at the basin level for economy-wide factor

   Capital level at time t:

$$KD_{i,r,f,a,t} = k_{i,r,a,t} \times \left(\frac{FPVA_{i,r,a,t} \times \alpha_{i,r,f,a,t}^i}{(1 + FWFDIST_{i,r,f,a,t})WF_{f,t}} \right)^{\sigma_{i,r,a}^i} \times FQA_{i,r,a,t} \times R_{i,r,a,t}$$

$$(4.19)$$

Land use level at time t:

$$NM_{i,r,f,a,t} = \eta_{i,r,a,t}$$
$$\times \left[\left(\frac{FPVA_{i,r,a} \times \alpha^i_{i,r,f,a}}{(1 + FWFDIST_{i,r,f,a})WF_f} \right)^{\sigma^i_{i,r,a}} \times FQA_{i,r,a} \div R_{i,r,a,t} - K_{i,r,f,a,t-1} \right]$$

$$(4.20)$$

Water resource consumption at time t:

$$WM_{i,r,f,a,t} = \omega_{i,r,a,t}$$
$$\times \left[\left(\frac{FPVA_{i,r,a} \times \alpha^i_{i,r,f,a}}{(1 + FWFDIST_{i,r,f,a})WF_f} \right)^{\sigma^i_{i,r,a}} \times FQA_{i,r,a} \div R_{i,r,a,t} - K_{i,r,f,a,t-1} \right]$$

$$(4.21)$$

where

$R_{i,r,a,t}$ return to capital at time t at the basin level
$k_{i,r,a,t}$ household income share of capital in sector i at the basin level
$\eta_{i,r,a,t}$ land use share of capital in sector i at the basin level
$\omega_{i,r,a,t}$ water resource consumption share of capital in sector i at the basin level

2. Sectoral demand at the basin level for economy-wide factor

$$FQF_{i,r,f,a} = (\Lambda^i_{i,r,a})^{\sigma^i_{i,r,a}-1} \times \left(\frac{FPVA_{i,r,a} \times \alpha^i_{i,r,f,a}}{(1 + FWFDIST_{i,r,f,a})WF_f} \right)^{\sigma^i_{i,r,a}} \times FQA_{i,r,a} \quad (4.22)$$

where

$FQF_{i,r,f,a}$ demand for factor, including water & land, by sector at the basin level

$FQA_{i,r,a}$,   production activity quantity at the basin level

$FWFDIST_{i,r,f,a}$ factor market distortion variables, including differences between water and land shadow prices and water and land market equilibrium price, at the basin level

$\alpha^i_{i,r,f,a}$ share parameter in CES value-added function at the perimeter level

3. CES value-added function at the basin level

$$FPVA_{i,r,a} = (\Lambda_{i,r,a}^i)^{-1} \times \left\{ \sum_{f \in EF} (\alpha_{i,r,f}^i)^{\sigma_{i,r,a}^i} \times [(1 + FWFDIST_{i,r,f,a})WF_f]^{1-\sigma_{i,r,a}^i} \right.$$
$$+ \sum_{f \in PF} (\alpha_{i,r,f}^i)^{\sigma_{i,r,a}^i} \times [(1 + FWFDIST_{i,r,f,a})FWF_{i,r,f}]^{1-\sigma_{i,r,a}^i}$$
$$\left. + \sum_{f \in WAT} (\alpha_{i,r,f}^i)^{\sigma_{i,r,a}^i} \times [1 + twa_{i,r,a} + (1 + FWFDIST_{i,r,f,a})FWF_{i,r,f}]^{1-\sigma_{i,r,a}^i} \right\}^{\frac{1}{(1-\sigma_{i,r,a}^i)}}$$

(4.23)

where

$FPVA_{i,r,a}$	value-added price at the basin level
$FWF_{i,r,f}$	factor price, including water and land market equilibrium price, at the basin level
$\Lambda_{i,r,a}^i$	shift parameter in CES value-added function at the basin level
$\alpha_{i,r,f}^i$	share parameter in CES activity composite function from perimeter to regional-level aggregation
$\sigma_{i,r,a}^i$	elasticity of substitution between factor inputs in CES value-added function at the perimeter level
$twa_{i,r,a}$	government water charge rate

4. Relationship between value-added and activity prices at the basin level

$$(1 - fta_{i,r,a})FPA_{i,r,a} \times FQA_{i,r,a} = FPVA_{i,r,a} \times FQA_{i,r,a} + \sum_c [fio_{i,r,c,a} \times PQ_c] \times FQA_{i,r,a}$$
$$+ \sum_c [FQINT\_G \times PQ_c]$$

(4.24)

where

$FPA_{i,r,a}$	production activity price at the basin level
$FQINT\_G$	energy demand for intermediate input in ground water production function at the perimeter level,

5. CES composite function between national- and regional-level activity prices

$$PA_a = (\Lambda_a^r)^{-1} \times \left[\sum_r (\alpha_{r,a}^{ro^r}) \times RPA_{r,a}^{1-\sigma_a^r} \right]^{\frac{1}{(1-\sigma_a^r)}}$$

(4.25)

where
Λ_a^r shift parameter in CES activity composite function from regional to national level aggregation

$RPA_{r,a}$ production activity price at the regional level

6. CES composite function between national- and regional-level activity prices

$$RPA_{r,a} = (\Lambda_{r,a}^i)^{-1} \times \left[\sum_i (\alpha_{i,r,a}^{i\,\sigma_{r,a}^i}) \times FPA_{i,r,a}^{1-\sigma_{r,a}^i} \right]^{\frac{1}{(1-\sigma_{r,a}^i)}} \tag{4.26}$$

where
$\Lambda_{r,a}^i$ shift parameter in CES activity composite function from basin to regional-level aggregation

$\sigma_{r,a}^i$ elasticity of substitution between basin-level output in CES composite activity function for the region

7. FOC for CES composite function from regional to national level activity aggregation

$$RQA_{r,a} = (\Lambda_a^r)^{\sigma_a^r-1} \times \left(\frac{PA_a \times \alpha_{r,a}^r}{RPA_{r,a}} \right)^{\sigma_a^r} \times QA_a \tag{4.27}$$

where
$RQA_{r,a}$ production activity quantity at the regional level

σ_a^r elasticity of substitution between regional level output in CES composite activity function for the country

8. FOC for CES composite function from basin to regional level activity aggregation

$$FQA_{r,a} = (\Lambda_{r,a}^i)^{\sigma_{r,a}^i-1} \times \left(\frac{RPA_{r,a} \times \alpha_{i,r,a}^i}{FPA_{i,r,a}} \right)^{\sigma_{r,a}^i} \times RQA_{r,a}. \tag{4.28}$$

4.2.4 Water Demand and Supply

1. Demand at the basin level for irrigation water

   (If irrigated water quantity is given by the government, this equation will give us water shadow prices)

$$FQF_{i,r,water} = \left(\Lambda_{i,r,a}^{i}\right)^{\sigma_{i,r,a}^{i}-1} \times FQA_{i,r,a}$$

$$\times \left(\frac{FPVA_{i,r,a} \times \alpha_{i,r,water,a}^{i}}{twa_{i,r,a} + (1 + FWFDIST_{i,r,water,a})FWF_{i,r,water}}\right)^{\sigma_{i,r,a}^{i}} \quad (4.29)$$

where

$FWF_{i,r,water}$ water price, including water and land market equilibrium price, at
the basin level

2. Urban water demand

$$WatQurb = shwat_{urb} \times QS_{elec\_wat} \quad (4.30)$$

where

$WatQurb$ ground water urban demand
$shwat_{urb}$ coefficient share of each basin in the total water supply

3. Total water demand at the basin level

(Given the constraint on total water supply, this equation gives us ground water
demand at the basin level)

$$FQF\_Tot_{i,r,a} = FQF_{i,r,water,a} + FQF\_G_{i,r,a} \quad (4.31)$$

where $FQF\_Tot_{i,r,a}$, total water demand at the basin level

4. Rural irrigation water supply at the basin level

(Given total water as an exogenous variable, increased urban demand for water
will reduce water availability for irrigation)

$$FQFS_{i,r,water} = shwat_{i,r}(WatQtot - WatQurb) \quad (4.32)$$

where

$FQFS_{i,r,water}$ supply of factor, including irrigated water, at the basin level
$shwat_{i,r}$ coefficient share of each perimeter in the total water supply
$WatQtot$ total water supply

5. Equations Transferring Activity into Commodity

$$QXAC_{a,c} = \vartheta_{a,c}QA_a \quad (4.33)$$

where $QXAC_{a,c}$, output from activity to commodity

$$PX_c = (\Lambda_c^c)^{-1} \times \left(\sum_a \delta_{a,c}^{a \, \sigma_c^c} \times PXAC_{a,c}^{1-\sigma_c^c} \right)^{\frac{1}{(1-\sigma_c^c)}} \tag{4.34}$$

where

Λ_c^c shift parameter in the CES function for transferring activities into commodity

$\delta_{a,c}^a$ share parameters in CES function for transferring activities into commodity

σ_c^c elasticity of substitution between activities in CES function for commodity

$$QXAC_{a,c} = (\Lambda_c^c)^{\sigma_c^c-1} \times \left(\frac{\delta_{a,c} \times PX_c}{PXAC_{a,c}} \right)^{\sigma_c^c} \times QX_c \tag{4.35}$$

where $\delta_{a,c}$, share parameters in CES function for transferring activities into commodity.

4.2.5 Imports and Exports

1. Armington demand function

$$QQ_c = \sum_{row} \left[\Lambda_{a,c}^{tr} \times \left[\left(\delta_{row,c} \times QM_{row,c}^{\frac{1-\sigma_c}{\sigma_c}} \right) + \left(\left(1 - \delta_{row,c}\right) \times QD_{row,c}^{\frac{1-\sigma_c}{\sigma_c}} \right) \right]^{\frac{\sigma_c}{(1-\sigma_c)}} \right] \tag{4.36}$$

where

$\Lambda_{a,c}^{tr}$ shift parameter in the Armington function

$\delta_{row,c}$ share parameters in CES function for transferring activities into commodity

2. Demand for import goods

$$QM_{row,c} = (\Lambda_{a,c}^{tr})^{-1} \times \left[\delta_{row,c}^{-\frac{\sigma_c}{(1-\sigma_c)}} \times \left(1 + \delta_{row,c}^m \times \left(\frac{PQ_c}{PM_{row,c}} \right)^{\sigma_c-1} \right)^{-\frac{\sigma_c}{(1-\sigma_c)}} \right]$$
$$\times QQ_{c,row} \tag{4.37}$$

where

$\delta^m_{row,c}$ share parameters in the CET function for imports

3. Demand for domestically produced goods

$$QD_{c,row} = (A^{tr}_{a,c})^{-1} \times (1 - \delta_{row,c})^{-\frac{\sigma_c}{(1-\sigma_c)}} \times \left(1 + \delta^e_{c,row} \times \left(\frac{PD_{row,c}}{PQ_c}\right)^{\sigma_c - 1}\right)$$
$$\times QQ_{c,row} \tag{4.38}$$

where

$\delta^e_{c,row}$ share parameters in the CET function for exports

4. Spillover goods

$$QX_c = \sum_{row} \left[A^e_{c,a} \times \left[\left(\delta^e_{c,row} \times QE_c^{-\frac{1+\sigma^e_c}{\sigma^e_c}} \right) + \left(\left(1 - \delta^e_{c,row}\right) QD_c^{-\frac{1+\sigma^e_c}{\sigma^e_c}} \right) \right]^{-\frac{\sigma^e_c}{1+\sigma^e_c}} \right] \tag{4.39}$$

$$QE_{c,row} = (A^e_{c,a})^{-1} \times \left[\delta^e_{row,c}^{-\frac{\sigma^e_c}{1+\sigma^e_c}} + \left(1 + \delta^e_{c,row} \times \left(\frac{PX_c}{PE_{c,row}}\right)^{-(1+\sigma^e_c)}\right)^{\frac{\sigma^e_c}{1+\sigma^e_c}} \right]$$
$$\times QX_{c,row} \tag{4.40}$$

$$PX_c = (A^c_c)^{-1} \times \left(\sum_a \delta^{a\,\sigma^c_c}_{a,c} \times PXAC^{1-\sigma^c_c}_{a,c} \right)^{\frac{1}{(1-\sigma^c_c)}} \tag{4.41}$$

where

PQ_c	composite commodity prices, $c \in C$
$PE_{c,row}$	export prices with margins and subsidies
$PM_{row,c}$	import prices with margins and tariffs
PX_c	aggregated producer price, $c \in C$
$QE_{c,row}$	the amount of export commodity
δ^e_c	CET elasticity of substitution between exports and domestically sold goods

5. Supply to domestic markets

$$
QD_{c,row} = (\Lambda_{c,a}^e)^{-1} \times \left[(1 - \delta_{row,c}^e)^{-\frac{\sigma_c^e}{1+\sigma_c^e}} + \left(1 + \delta_{c,row} \times \left(\frac{PX_c}{PE_{c,row}} \right)^{-(1+\sigma_c^e)} \right)^{\frac{\sigma_c^e}{1+\sigma_c^e}} \right]
$$
$$
\times\; QX_{c,row}
$$

(4.42)

4.2.6 Incomes and Demands

1. Economy-wide factor income

$$
\begin{aligned}
YF_{f\in EF} = & \sum_{a\notin APIR} (1 + WFDIST_{f,a}) \times WF_f \times QF_{f,a} \\
& + \sum_{a\in APIR} \sum_{i,r} (1 + FWFDIST_{i,r,f,a}) \times WF_f \times FQF_{i,r,f,a}
\end{aligned}
$$

(4.43)

where
$EF \subset F$ economy-wide factor
YF_f factor income

2. Basin-specific factor income

$$
YF_{f\in PF} = \sum_{a\in APIR} \sum_{i,r} (1 + FWFDIST_{i,r,f,a}) \times FWF_{i,r,f} \times FQF_{i,r,f,a}
$$

(4.44)

where
$PF \subset F$ primary factors

3. Water income

$$YF_{f \in EF} = \sum_{a \notin APIR} (1 + WFDIST_{f,a}) \times WF_f \times QF_{f,a}$$

$$+ \sum_{a \in APIR} \sum_{i,r} (1 + FWFDIST_{i,r,f,a}) \times WF_{i,r,f} \times FQF\_Tot_{i,r,f,a}$$

$$- \left(\sum_{a \in APIR} \sum_{i,r} twa_{i,r,a} \right) \times FQF_{i,r,f,a}$$

(4.45)

4. Factor income distributed to households

$$YIF_{h,f} = shif_{h,f} \times YF_f (1 - tf_f)$$ (4.46)

where

INS	set of institutions such as households, government, other domestic area and foreign trading partner countries
$H \subset INS$	set of households
$YIF_{h,f}$	factor income for different households
$shif_{h,f}$	initial distribution of factor income across households
tf_f	tax rate on factor income

5. Household income

$$YI_h = \sum_f YIF_{h,f} + \sum_{ins} trnsfr_{ins,h}$$ (4.47)

6. Household demand

$$QH_{c,h} = \frac{\beta_{c,h} \times \left(YI_h \times (1 - SADJ \times mps_h) - \sum_{c'} PQ_{c'} \times \gamma_{c',h} \right)}{PQ_c} + \gamma_{c,h}$$ (4.48)

where

QH_{ch}	household demand
SADJ	savings adjustment factors
mps_h	household saving rate
$\beta_{c,h}$	share parameter in household's demand function
$\gamma_{c,h}$	subsistence parameter in the Stone–Geary utility function

7. Government revenue

$$
\begin{aligned}
YG_r = \sum_{a \notin APIR} ta_a \times PA_a \times QA_a + \sum_{a \in APIR} \sum_{i,r} fta_{i,r,a} FPA_{i,r,a} FQA_{i,r,a} \\
+ \sum_{row} \sum_{c} [tm_{row,c} \times EXR \times pwm_{row,c} \times QM_{row,c}] \\
+ \sum_{c} \sum_{row} [te_{c,row} \times EXR \times pwm_{c,row} \times QE_{c,row}] \\
+ \sum_{c} [tq_c \times PQ_{i,r} \times QQ_c] + \sum_{f} tf_f \times YF_f \\
+ \sum_{i,r} twa_{i,r} \times FQF_{i,r,water} \\
+ \sum_{row} (trnsfr_{row,gov't,row} - trnsfr_{gov't,row}) \times EXR
\end{aligned}
\tag{4.49}
$$

where

YG_r	government income
ta_a	tax rate on production activity
$fta_{ir,a}$	government water charge rate
$tm_{row,c}$	tariff rate on imports
$te_{c,row}$	tax rate on exports
$trnsfr_{row,gov't,row}$	transfers between institutions

8. Government spending on commodities

$$
PQ_c \times QG_c = GADJ \times \overline{qg_c}
\tag{4.50}
$$

where

QG_c	government demand
$GADJ$	government demand scaling factors
$fta_{i,r,a}$	government water charge rate
$\overline{qg_c}$	initial value of government spending on commodity

9. Government total expenditure

$$
EG = \sum_{c} PQ_c \times QG_c + \sum_{h} trnsfr_{govt,h}
\tag{4.51}
$$

where

EG	government expenditure
$trnsfr_{govt,h}$	transfer from institute to household

10. Government budget surplus

$$GSAV = YG - EG \qquad (4.52)$$

where
$GSAV$ government savings

11. Investment demand

$$PQ_c \times QINV_c = shinv_c \times totvinv \qquad (4.53)$$

where
$QINV_c$ investment demand
$shinv_c$ initial share of investment on commodity
$totvinv$ total investment

12. National-level savings

$$SAVINGS_c = \sum_h SADJ \times mos_h \times YI_h + GSAV + tfsav \times EXR \qquad (4.54)$$

where
$SAVINGS_c$ total savings
$tfsav$ total trade deficits

13. Equilibrium Conditions

Commodity markets

$$QQ_c = QINT_c + \sum_h QH_{c,h} + QG_c + QINV_c. \qquad (4.55)$$

4.2.7 Factor Markets

1. Markets for the factors employed in non-APIR sectors only

$$\sum_{a \notin APIR} QF_{f,a} = FS_f \qquad (4.56)$$

2. Segmented markets at the basin level for the factors employed in APIR sectors only

$$\sum_{a \in APIR} FQF_{i,r,f,a} = FFS_{i,r,f} \tag{4.57}$$

3. Markets for the economy-wide factors

$$\sum_{a \notin APIR} QF_{f,a} + \sum_{a \in APIR} \sum_{i,r} FQF_{i,r,f,a} = FS_f \tag{4.58}$$

4. Foreign savings

$$FSAV_{row} = \sum_{c} \left(pwm_{c,row} \times QM_{c,row} - pwe_{row,c} \times QE_{row,c} \right)$$
$$+ \sum_{ins} \left(trnsfr_{ins,row} - trnsfr_{row,ins} \right) \tag{4.59}$$

where
$FSAV_{row}$ trade deficits
$trnsfr_{ins,row}$ transfer payment of institution from domestic to foreign
$trnsfr_{row,ins}$ transfer payment of institution from foreign to domestic

$$TFSAV = \sum_{row} FSAV_{row} \tag{4.60}$$

4.3 Model Notations

Sets

A	Activities
$APIR \subset A$	Activities inside water districts (basin)
C	Commodities
F	Factors, including water and land, employed in activities
$EF \subset F$	Economy-wide factor
$WF \subset F$	*basin-specific factor*
$WAT \subset F$	Water
$LND \subset F$	Land
INS	Institutions such as households, government, other domestic area and foreign trading partner countries
$H \subset INS$	Households

(continued)

(continued)

$DT \subset INS$	other domestic area
$ROW \subset INS$	Foreign trading partner countries
R	Water administrative regions
I	Water administrative districts (basin)

Variables

Exogenous Variables at the National Level

$PWE_{c,row}$	Import prices (fob.), $c \in C$, row $\in$ ROW
$PWM_{row,c}$	Import prices (fob.), $c \in C$, row $\in$ ROW
EXR	Foreign exchange rate, row $\in$ ROW
$TFSAV$	Total trade deficits
$INVEST$	Total investment value
$WatQtot$	Total water supply
$LndQtot$	Total land supply

Endogenous Prices at the National Level

L_i	production factor of labor in sector i
K_i	production factor of capital in sector i
N_i	production factor of land use in sector i
W_i	production factor of water resource consumption in sector i
CPI	Consumer price index
PA_a	Production activity price (unit gross revenue), $a \in A$
PVA_a	Value-added price, $a \in A$
$PXAC_{a,c}$	Price from activity to commodity $a \in A$, $c \in C$
PX_c	Aggregated producer price, $c \in C$
PD_c	Price for commodity produced and sold domestically, $c \in C$
PQ_c	Composite commodity prices, $c \in C$
$PE_{c,row}$	Export prices with margins and subsidies, $c \in C$, row $\in$ ROW
$PM_{row,c}$	Import prices with margins and tariffs, $c \in C$, row $\in$ ROW
WF_f	Factor price, including water price and land rent, $f \in F$
$WFDIST_{f,a}$	Factor market distortion variables, $f \in F$, $a \in A$

Production-related Endogenous Variables at the National Level

QA_a	Output from activity to commodity, $a \in A$, $c \in C$
QX_c	Commodity output, $c \in C$
QD_c	Commodity produced and sold domestically, $c \in C$
$QINT_c$	Aggregated intermediate demand, $c \in C$
$QF_{f,a}$	Demand for factor by sector, $f \in F$, $a \in A$
QFS_f	Factor supply, $f \in F$

Demand-related Endogenous Variables at the National Level

QQ_c	Armington composite commodity, $c \in C$
$QE_{c,row}$	Exports, $c \in C$, row $\in$ ROW
$QM_{row,c}$	Imports, $c \in C$, row $\in$ ROW
QM_c	Imports, $c \in C$
$QH_{c,h}$	Household demand, $c \in C$, $h \in H$
QG_c	Government demand, $c \in C$
$QINV_c$	Investment demand, $c \in C$
$WatQurb$	Ground water urban demand
$WatQrurb$	Total rural water demand in irrigated areas

(continued)

(continued)

$LndQind$	The industry land demand
$LndQcrop$	The cropland demand

Aggregated and Macroeconomic Endogenous Variables at at the National Level

EG	Government expenditure
YG	Government income
$GSAV$	Government savings
$GADJ$	Government demand scaling factors
YF_f	Factor income, $f \in F$
$YIFh_{,f}$	Factor income for different households, $h \in H, f \in F$
YIh	Household income, $h \in H$
$SADJ$	Savings adjustment factors
$SAVINGS$	Total savings
$FSAV_{row}$	Trade deficits, $row \in ROW$

Endogenous Prices at the Subnational Level

$FPA_{i,r,a}$	Production activity price at the basin level, $i \in I, r \in R, a \in APIR$
$RPA_{r,a}$	Production activity price at the regional level, $r \in R, a \in APIR$
$FPVA_{i,r,a}$	Value-added price at the basin level, $i \in I, r \in R, a \in APIR$
$FWF_{i,r,f}$	Factor price, including water and land market equilibrium price, at the basin level, $r \in R, f \in F$
$FWFDIST_{i,r,f,a}$	Factor market distortion variables, including differences between water & land shadow prices and water & land market equilibrium price, at the basin level, $i \in I$, $r \in R, f \in F, a \in APIR$

Production-related Endogenous Variables at the Subnational Level

$FQA_{i,r,a}$	Production activity quantity at the basin level, $i \in I, r \in R, a \in APIR$
$RQA_{r,a}$	Production activity quantity at the regional level, $r \in R, a \in APIR$
$FQF_{i,r,f,a}$	Demand for factor, including water, by sector at the basin level, $i \in I, r \in R, f \in F$, $a \in APIR$
$FQF_{Si,r,f}$	Factor, including irrigated water, supply at the basin level, $i \in I, r \in R, f \in F$
$FQFS\_G_{i,r}$	Ground water supply at the basin level, $i \in I, r \in R, f \in F, a \in APIR$
$FQF\_Tot_{i,r}$	Total water demand at the basin level, $i \in I, r \in R, f \in F, a \in APIR$
$QINT\_G_{i,r,c,a}$	Energy demand for intermediate input in ground water production function at the basin level, $i \in I, r \in R, a \in APIR$

Parameters

Assumed Parameters in Equations for the National Economy

A_i	shift parameter of activity in sector i
δ_i	Elasticity of labor input for output in sector i
κ_i	Elasticity of capital input for output in sector i
η_i	Elasticity of land use for output in sector i
ω_i	Elasticity of water resource consumption for output in sector i
ρ_i	Elasticity of substitution between factor inputs in production function
σ_a	Elasticity of substitution between factor inputs in CES value-added function
α_c^c	Elasticity of substitution between activities in CES function for commodity
$\alpha_{row,c}^a$	Armington elasticity of substitution between domestic and import goods
$\alpha_{c,row}^e$	CET elasticity of substitution between exports and domestically sold goods

Computed Parameters in Equations for the National Economy

$\beta_{c,h}$	Share parameter in household's demand function

(continued)

(continued)

$\gamma_{c,h}$	Subsistence parameter in the Stone–Geary utility function
$\alpha_{f,a}^a$	Share parameter in the CES value-added function
Λ_a	Shift parameter in the CES value-added function
$io_{c,a}$	Input–output coefficient
$\delta_{a,c}$	Share parameters in CES function for transferring activities into commodity
$\delta_{row,c}^m$	Share parameters in the Armington function for imports
$\delta_{c,row}^e$	Share parameters in the CET function for exports
$\Lambda_{a,c}^a$	Shift parameter in the CES function for transferring activities into commodity
$\Lambda_{a,c}^{tr}$	Shift parameter in the Armington function
$\Lambda_{c,row}^e$	Shift parameter in the CET function

Parameters in Equations for the Subnational Economy

σ_a^r	Elasticity of substitution between regional-level output in CES composite activity function for the country
$\sigma_{r,a}^i$	Elasticity of substitution between basin-level output in CES composite activity function for the region
$a_{r,a}^r$	Share parameter in CES activity composite function from region to national-level aggregation
$a_{i,r,a}^i$	Share parameter in CES activity composite function from basin to regional-level aggregation
Λ_a^r	Shift parameter in CES activity composite function from region to national-level aggregation
$\Lambda_{r,a}^i$	Shift parameter in CES activity composite function from basin to regional-level aggregation
$\delta_{i,r,a}^i$	Elasticity of substitution between factor inputs in CES value-added function at the basin level
$a_{i,r,f,a}^i$	Share parameter in CES value-added function at the basin level
$\Lambda_{i,r,a}^i$	Shift parameter in CES value-added function at the basin level
$fio_{i,r,c,a}$	Input–output coefficient at the basin level
$\vartheta_{a,c}$	Yield of output c per unit of activity a

Other Computed Parameters

ta_a	Tax rate on production activity
ta_a	Tax rate on consumption
tec	Tax rate on exports
tmc	Tariff rate on imports
tf_f	Tax rate on factor income
$twa_{i,r,a}$	Government water charge rate
$fta_{i,r,a}$	Government water charge rate
$trnsfri_{ns,ins'}$	Transfers between institutions
$shif_{h,f}$	Initial distribution of factor income across households
mps_h	Household saving rate
$\bar{q}\bar{g}_c$	Initial value of government spending on commodity
$shinv_c$	Initial share of investment on commodity
$cwts_c$	Initial share of investment on commodity
$shwati_{r,a}$	Coefficient share of each basin in the total water supply
$shwat_{urb}$	Coefficient share of each basin in the total water supply

Reference

Ghadimi, H. (2007). CGE Modeling and Applications: A Short Course In. Morgantown, West Virginia. http://rri.wvu.edu/CGECourse/index.htm

Chapter 5
Implementation

CGE models (Fig. 5.1) are referred to as computable because they can be applied to economic data. Data for the SAM is collected and then adjusted and balanced so that total receipts are equal to total outlays for each account. The SAM data represents the so-called benchmark general equilibrium, along with specific assumptions regarding utility and production functions to show one equilibrium solution of the economic model (Deng 2011). An integrated CGE model will usually include a set of water and land accounts that accompany the SAM, which represent water and land use by industry and final demand sectors at the equilibrium solution.

Since the benchmark is considered to represent an equilibrium solution, once specific functional forms are chosen, the benchmark data is used to calibrate the parameter values for the functional forms. Depending on the functional forms chosen for producers and consumers, some parameter values will not be supplied by the calibration and will have to be supplied exogenously. Values are either taken from the literatures or chosen using the modeler's best judgment.

After calibration, the model is checked to see if it correctly replicates the baseline data in the SAM. When it is established that the baseline data can be replicated, the model is "shocked". For example, an increase in export demand may be imposed exogenously or a tax may be eliminated. The model is solved once again to find the "counterfactual" equilibrium set of prices and quantities for all sectors. These results can then be compared to the base solution or other counterfactual scenarios (Deng 2011).

To illustrate the type of comparative policy analysis that can be carried out with an integrated CGE and what outputs from such a model look like, we elaborate results from Qureshi et al. (2012), who use the static version of the Multi-regional model TERM-H2O to examine how water resources and their prices are affected by an increase in the urban population and decreases in water availability. Four scenarios are modeled. The baseline year is shocked with population growth and decreasing water availability under four different policy alternatives:

X. Deng et al., *Integrated River Basin Management*,
SpringerBriefs in Environmental Science,
DOI: 10.1007/978-3-662-43466-6_5, © The Author(s) 2014

Fig. 5.1 Diagram of CGE modeling process

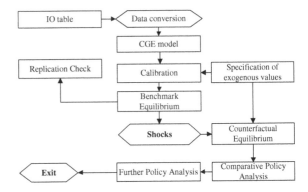

(a) Business as usual: no water trades between regions are allowed and no new water sources are developed.
(b) Water trading between rural and urban areas.
(c) Water trading is allowed and a "new" water source is built (perhaps a desalinization plant).
(d) Scenario c is modified by allowing labor mobility between regions.

For each scenario, Qureshi et al. (2012) report aggregate consumption, real gross regional product, aggregate employment for each region, water use by sector, water price (use charges) and shadow prices of water for each region. They find that without new water sources or water trading, regional cities will face as much as an eightfold increase in the shadow price of a kiloliter of water. The "business as usual" scenario above will result in the lowest level of aggregate consumption. Water trade between urban and rural regions will reduce production in water intensive crops as the shadow price of water increases in rural areas and decreases in urban areas to equilibrate urban and rural water prices. Providing new supplies, even after accounting for infrastructure costs, reduces the economic impact on rural areas.

Because the CGE model is a representation of the entire economy, the output from the model gives a complete set of market-clearing prices and quantities in the product and factor markets. Thus almost any economic variable of interest can be compared to the baseline: GDP, employment levels by sector, aggregate consumption, water use by sector, water prices and shadow values, and so forth. Especially important for the integrated CGE model is that an explicit measure of welfare, the equivalent variation, can be calculated from the results so that the change in welfare for different simulations can be calculated. This serves as a shadow price for water in some models where simulations change the quantity of water available to the economy.

5.1 Interactive Model Building Environment

GEMPACK (General Equilibrium Modeling Package) is a suite of economic modeling software designed for building and solving applied general equilibrium models. It can handle a wide range of economic behaviors and contains powerful capabilities for solving inter-temporal models. Therefore, the software is suggested to apply to the construction of CGE model in the research. GEMPACK calculates accurate solutions of an economic model, starting from an algebraic representation of the model equations. These equations can be written as levels equations, linearized equations or a mixture of these two.

The complete ranges of GEMPACK features are available only on PCs running Microsoft Windows XP or later version operating system. Nevertheless GEMPACK will run (with some limitations) on other operating systems, such as MacOS or UNIX. GEMPACK supports both 32-bit and 64-bit versions of Windows XP, Windows Vista and Windows 7. The 64-bit versions are more suitable for large modeling tasks, particularly if you wish to exploit the parallel-processing capability of modern PCs. The GEMPACK Web site is at: http://www.monash.edu.au/policy/gempack.htm. This contains up-to-date information about GEMPACK, including information about different versions, prices, updates, courses and bug fixes. We encourage GEMPACK users to visit this site regularly.

Instructions for installation of GEMPACK Release 11 on a PC which is running Windows are provided below. To install GEMPACK Release 10 (or earlier) please refers to the install documents that accompanied the earlier Release.

Some parts of the install procedure differ between the Executable-Image and the Source-Code versions of GEMPACK—the text below will indicate these differences. All components of GEMPACK are contained in a single install package, which you might download or receive on a CD.

The install package for Executable-Image GEMPACK might have a name like gpei-11.0-000-install.exe. While, the install package for Source-Code GEMPACK might have a name like gpsc-11.0-000-install.exe. Both the packages will install Windows (GUI) programs such as ViewHAR, ViewSOL, TABmate, AnalyseGE, WinGEM and RunGEM. The Executable-Image package will also install a number of vital command-line programs such as TABLO.EXE and GEMSIM.EXE. The Source-Code package instead installs Fortran source code files for these programs: during installation these sources are compiled to produce TABLO.EXE and GEMSIM.EXE.

The ideal plan is to install GEMPACK in a folder C:\GP to which the user has read/write/modify access. If you have another or earlier release of GEMPACK already installed in C:\GP, you should probably leave it on your hard disk until you have successfully installed and tested Release 11.0 (in case an unexpected problem occurs). You should rename the directory containing the previous release, so you can install Release 11.0 of GEMPACK in C:\GP. For example, if you currently have Release 9.0 GEMPACK in C:\GP, rename that directory to, say, C:\GP90, and install Release 11.0 into C:\GP. If you use the same GEMPACK directory as before, other GEMPACK-related Windows programs such as RunDynam and RunGTAP will automatically use the latest version of GEMPACK (Fig. 5.2). It's best if both the GEMPACK programs and the user's model files are stored on a local hard drive (not a network drive).

In the main stages of the GEMPACK, the first and largest task, the specification of the model's equations using the TABLO language, has been described at length in the previous sections. This material is contained in the ***.TAB file (at top left of the Fig. 5.2).

Equation E_x1_s # Demands for Commodity Composites #

(All,c,COM)(All,i,IND)   x1_s(c,i) - {a1_s(c,i) + a1tot(i)} = x1tot(i);

The model as described so far has too many equations and variables for efficient solution. Their numbers are reduced by instructing the TABLO program to omit specified variables from the system. This option is useful for variables which will be exogenous and unshocked (zero percentage change). Normally it allows us to dispense with the bulk of the technical change terms. Of course, the particular selection of omitted variables will alter in accordance with the model simulations to be undertaken. Substitute out specified variables using specified equations can results in fewer but more complex equations. Typically we use this method to eliminate multi-dimensional matrix variables which are defined by simple equations. For example, the equation that appears in the TABLO Input file, can be used to substitute out variable x1_s. In fact the names of the MODEL equations are chosen to suggest which variable each equation could eliminate. The variables for omission and the equation-variable pairs for substitution are listed in a second, instruction, file: ***.STI.

The TABLO program converts the TAB and STI files into a Fortran source file, ***.FOR, which contains the model-specific code needed for a solution program. The Fortran compiler combines ***.FOR with other, general-purpose, code to produce the executable program ***.EXE, which can be used to solve the model specified by the user in the TAB and STI files (Fig. 5.3).

Simulations are conducted using ***.EXE. Its input is a data file, containing input-output data and behavioral parameters. This data file contains all necessary information about the initial equilibrium user input from a text (CMF) file, which specifies:

- which variables are to be exogenous;
- shocks to some exogenous variables;
- into how many steps the computation is divided;
- the names of input and output files, and other details of the solution process.

Fig. 5.2 The flow of CGE model using GEMPACK software

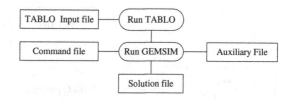

Fig. 5.3 Building a model-specific EXE file

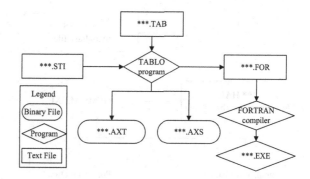

Each simulation produces an SL4 (Solution) file. The SL4 file has a binary format: it may be viewed with the non-model-specific Windows program ViewSOL.

Figure 5.4 shows a variation on the processes depicted in Fig. 5.3. This time, TABLO has produced a GSS file. Unlike the FOR file of Fig. 5.3, the GSS file need not be compiled: it is interpreted directly by the standard program GEMSIM. The advantages of this approach are that no Fortran compiler is needed and that the required GEMPACK license is cheaper. The disadvantage is that larger models may solve only slowly, or may altogether exceed size limits built into GEMSIM. Both methods give the same numerical results.

5.2 Database of the Model

Table 5.1 is a schematic representation of the model's input–output database. It reveals the basic structure of the model. The column headings in the main part of the table (an absorption matrix) identify the following demanders:

- Domestic producers divided into I industries;
- Investors divided into I industries;
- A single representative household;
- An aggregate foreign purchaser of exports;
- Government demands;
- Changes in inventories.

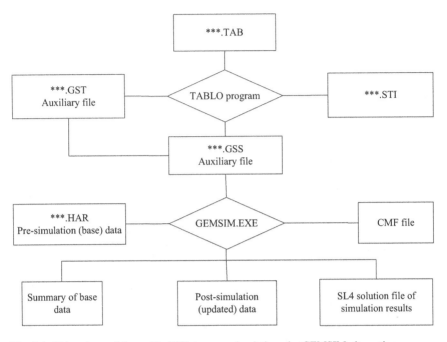

Fig. 5.4 Using the model-specific EXE to run a simulation, the GEMSIM alternative

	Joint production matrix
Size	I
C	MAKE

	Import duty
Size	1
C	V0TAR

The entries in each column show the structure of the purchases made by the agents identified in the column heading. Each of the C commodity types identified in the model can be obtained locally or imported from overseas. The source-specific commodities are used by industries as inputs to current production and capital formation, consumed by households and governments, which are exported, or are added to or subtracted from inventories. Only domestically produced goods appear in the export column. M of the domestically produced goods are used as margins services (wholesale and retail trade, and transport) which are required to transfer commodities from their sources to their users. Commodity taxes are payable on the purchases. As well as intermediate inputs, current production requires inputs of four categories of primary factors: water (divided into three types, including groundwater, surface water and reclaimed water), labor, fixed capital, and agricultural land. Production taxes include output taxes or subsidies that are not user-specific. The 'other costs' category covers various miscellaneous taxes on firms, such as municipal taxes or charges.

Table 5.1 The model flows database

Size		Absorption matrix					
		1	2	3	4	5	6
		Producers	Investors	Household	Export	Government	Change in inventories
		1	1	1	1	1	1
Basic flows	CS	V1BAS	V2BAS	V3BAS	V4BAS	V5BAS	V6BAS
Margins	CSM	V1MAR	V2MAR	V3MAR	V4MAR	V5MAR	n/a
Taxes	CS	V1TAX	V2TAX	V3TAX	V4TAX	V5TAX	n/a
Labor	O	V1LAB	C = Number of commodities				
Capital	1	V1CAP	I = Number of industries				
Water	3	V1CAP					
Land	1	V1LND	S = 2: Domestic, imported				
Production tax	1	V1PTX	O = Number of occupation types				
Other costs	1	V1OCT	M = Number of commodities used as margins				

Each cell in the illustrative absorption matrix in Table 5.1 contains the name of the corresponding data matrix. For example, V2MAR is a 4-dimensional array showing the cost of M margins services on the flows of C goods, both domestically produced and imported (S), to I investors.

In principle, each industry is capable of producing any of the C commodity types. The MAKE matrix at the bottom of Table 5.1 shows the value of output of each commodity by each industry. Finally, tariffs on imports are assumed to be levied at rates which vary by commodity but not by user. The revenue obtained is represented by the tariff vector V0TAR.

Excerpt 1 of the TABLO Input file begins by defining logical names for input and output files. Initial data are stored in the BASEDATA input file. The

```
! Excerpt 1 of TABLO input file: !

! Files and sets !

File   BASEDATA   # Input data file #;

 (new) SUMMARY    # Output for summary and checking data #;


Set                                                                                    !Index!

  COM # Commodities# read elements from file BASEDATA header "COM";        ! c !

  SRC # Source of commodities # (dom,imp);                                 ! s !

  IND # Industries # read elements from file BASEDATA header "IND";        ! i !

  OCC # Occupations # read elements from file BASEDATA header "OCC";       ! o !

  MAR # Margin commodities # read elements from file BASEDATA header "MAR";! m !

  Subset MAR is subset of COM;

  Set NONMAR      # Non-margin commodities # = COM - MAR;                  ! n !
```

SUMMARY output file is used to store summary and diagnostic information. Note that BASEDATA and SUMMARY are logical names. The actual locations of these files (disk, folder, filename) are chosen by the model user.

The rest of Excerpt 1 defines sets: lists of descriptors for the components of vector variables. Set names appear in upper-case characters. For example, the first Set statement is to be read as defining a set named 'COM' which contains commodity descriptors. The elements of COM (a list of commodity names) are read from the input file (this allows the model to use databases with different numbers of sectors). By contrast, the two elements of the set SRC—dom and imp—are listed explicitly.

The commodity, industry, and occupational classifications of the regional version of MODEL described here are aggregates of the classifications used in the original version of ORANI, which had over 100 industries and commodities, and 8 labor occupations.

The industry classification differs slightly from the commodity classification. Both are listed in Table 5.1. In this aggregated version of the model, multi-production is confined to the first two industries, which produce the first three commodities. Each of the remaining industries produces a unique commodity. Labor is disaggregated into skill-based occupational categories described by the set OCC.

The central column of Table 5.1 lists the elements of the set COM which are read from file. GEMPACK uses the element names to label the rows and columns of results and data tables. The element names cannot be more than 12 letters long, nor contain spaces. The IND elements are the same as elements 2–23 of COM.

Elements of the set MAR are margins commodities, i.e., they are required to facilitate the flows of other commodities from producers (or importers) to users. Hence, the costs of margins services, together with indirect taxes, account for differences between basic prices (received by producers or importers) and purchasers' prices (paid by users).

TABLO does not prevent elements of two sets from sharing the same name; nor, in such a case, does it automatically infer any connection between the corresponding elements. The Subset statement which follows the definition of the set MAR is required for TABLO to realize that the two elements of MAR, Trade and Transport, are the same as the 18th and 19th elements of the set COM.

The statement for NONMAR defines that set as a complement. That is, NONMAR consists of all those elements of COM which are not in MAR. In this case TABLO is able to deduce that NONMAR must be a subset of COM.

5.3 The Percentage-Change Approach to Model Solution

Many of the model equations are non-linear—demands depend on price ratios, for example. However, following Johansen (1960), the model is solved by representing it as a series of linear equations relating percentage changes in model variables. This section explains how the linearized form can be used to generate exact solutions of the underlying, non-linear, equations, as well as to compute linear approximations to those solutions.[1]

A typical Applied General Equilibrium (AGE) model can be represented in the levels as:

$$\mathbf{F}(\mathbf{Y}, \mathbf{X}) = \mathbf{0} \tag{5.1}$$

where, $\mathbf{Y}$ is a vector of endogenous variables, $\mathbf{X}$ is a vector of exogenous variables and F is a system of non-linear functions. The problem is to compute $\mathbf{Y}$, given $\mathbf{X}$. Normally we cannot write $\mathbf{Y}$ as an explicit function of $\mathbf{X}$.

Several techniques have been devised for computing Y. The linearized approach starts by assuming that we already possess some solution to the system, $\{Y^0, X^0\}$, i.e.

$$\mathbf{F}(\mathbf{Y}^0, \mathbf{X}^0) = \mathbf{0} \tag{5.2}$$

Normally the initial solution $\{\mathbf{Y}^0,\ \mathbf{X}^0\}$ is drawn from historical data—we assume that our equation system was true for some point in the past. With conventional assumptions about the form of the F function it will be true that for small changes $\mathbf{dY}$ and $\mathbf{dX}$:

$$\mathbf{F_Y}(\mathbf{Y}, \mathbf{X})\mathbf{dY} + \mathbf{F_X}(\mathbf{Y}, \mathbf{X})\mathbf{dX} = \mathbf{0} \tag{5.3}$$

where $\mathbf{F_Y}$ and $\mathbf{F_X}$ are matrices of the derivatives of $\mathbf{F}$ with respect to $\mathbf{Y}$ and $\mathbf{X}$, evaluated at $\{\mathbf{Y_0}, \mathbf{X_0}\}$. For reasons explained below, we find it more convenient to express $\mathbf{dY}$ and $\mathbf{dX}$ as small percentage changes y and x. Thus y and x, some typical elements of y and x, are given by:

$$y = 100\mathbf{dY}/\mathbf{Y} \text{ and } x = 100\mathbf{dX}/\mathbf{X} \tag{5.4}$$

Correspondingly, we define:

$$\mathbf{G_Y}(\mathbf{Y}, \mathbf{X}) = \mathbf{F_Y}(\mathbf{Y}, \mathbf{X})\hat{\mathbf{Y}} \text{ and } \mathbf{G_X}(\mathbf{Y}, \mathbf{X}) = \mathbf{F_X}(\mathbf{Y}, \mathbf{X})\hat{\mathbf{X}} \tag{5.5}$$

where, $\hat{\mathbf{Y}}$ and $\hat{\mathbf{X}}$ are diagonal matrices. Hence the linearized system becomes:

$$\mathbf{G_Y}(\mathbf{Y}, \mathbf{X})\mathbf{y} + \mathbf{G_X}(\mathbf{Y}, \mathbf{X})\mathbf{x} = \mathbf{0} \tag{5.6}$$

Such systems are easy to solve for computers by using standard techniques of linear algebra. But they are accurate only for small changes in $\mathbf{Y}$ and $\mathbf{X}$. Otherwise, linearization error may occur. The error is illustrated by Fig. 5.5, which shows how some endogenous variable Y changes as an exogenous variable X moves from

---

[1] For a detailed treatment of the linearized approach to AGE modeling, see the Black Book. Chapter 3 contains information about Euler's method and multistep computations.

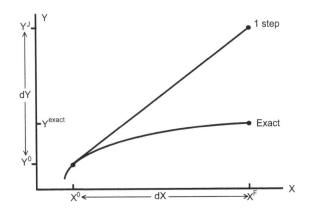

Fig. 5.5 Linearization error. *Note* Fig. 5.5 suggests that, the larger x is reached, the greater the proportional error in y would be gotten. This observation breaks large changes in X into a number of steps, as shown in Fig. 5.6. For each sub-change in X, we use the linear approximation to derive the consequent sub-change in Y. Then, using the new values of X and Y, we recomputed the coefficient matrices $\mathbf{G_Y}$ and $\mathbf{G_X}$. The process is repeated for each step. If we use 3 steps (see Fig. 5.6), the final value of Y, Y3, is closer to Yexact than was the Johansen estimate YJ. We can show, in fact, that given sensible restrictions on the derivatives of $\mathbf{F(Y, X)}$, we can obtain a solution as accurate as we like by dividing the process into sufficiently many steps

X_0 to X_F. Theoretically, non-linear relation between X and Y is shown as a curve. The linear, or first-order, approximation:

$$y = -\mathbf{G_Y}(\mathbf{Y}, \mathbf{X})^{-1}\mathbf{G_X}(\mathbf{Y}, \mathbf{X})x \tag{5.7}$$

leads to the Johansen estimate Y^J—an approximation to the actual series, Y^{exact}.

The technique illustrated in Fig. 5.6, known as the Euler method, is the simplest one in several related techniques for numerical integration—the process of using differential equations (change formulae) to move from one solution to another. GEMPACK offers the choice of several such techniques. Each of them requires inputting an initial solution $\{\mathbf{Y^0}, \mathbf{X^0}\}$ and formulating for the derivative matrices $\mathbf{G_Y}$ and $\mathbf{G_X}$ tends to the total percentage change in the exogenous variables, x. The levels functional form, $\mathbf{F(Y, X)}$, need not be specified, although it underlies $\mathbf{G_Y}$ and $\mathbf{G_X}$.

Furthermore, the accuracy of multistep solution techniques can be improved by extrapolation. Suppose the same experiments are repeated 4-step, 8-step and 16-step by Euler computations which yield the following estimates for the total percentage change in endogenous variable Y respectively:

y(4-step) = 4.5 %,
y(8-step) = 4.3 % (0.2 % less), and
y(16-step) = 4.2 % (0.1 % less).
Extrapolation suggests that the 32-step solution would be:
y(32-step) = 4.15 % (0.05 % less),
and that the exact solution would be:
y(∞-step) = 4.1 %.

Fig. 5.6 Multi-step
processes to reduce
linearization error

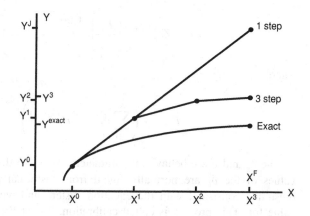

The extrapolated result requires 28 ($= 4 + 8 + 16$) steps to compute but would
normally be more accurate than that given by a single 28-step computation.
Alternatively, extrapolation enables us to obtain given accuracy with fewer steps.
As we noted above, each step of a multi-step solution requires: computation from
data of the percentage-change derivative matrices $\mathbf{G_Y}$ and $\mathbf{G_X}$; solution of the
linear system (6); and use of that solution to update the data $(\mathbf{X}, \mathbf{Y})$.

In practice, for typical AGE models, it is unnecessary, during a multistep
computation, to record values for every element in $\mathbf{X}$ and $\mathbf{Y}$. Instead, we can define
a set of data coefficients $\mathbf{V}$, which are functions of $\mathbf{X}$ and $\mathbf{Y}$, i.e., $\mathbf{V} = \mathbf{H}(\mathbf{X}, \mathbf{Y})$.
Most elements of $\mathbf{V}$ are simple cost or expenditure flows such as appear in input-
output tables. $\mathbf{G_Y}$ and $\mathbf{G_X}$ turn out to be simple functions of $\mathbf{V}$; often indeed
identical to elements of $\mathbf{V}$. After each small change, $\mathbf{V}$ is updated using the
formula $\mathbf{v} = \mathbf{H_Y}(\mathbf{X}, \mathbf{Y})\mathbf{y} + \mathbf{H_X}(\mathbf{X}, \mathbf{Y})\mathbf{x}$. The advantages of storing $\mathbf{V}$, rather than
$\mathbf{X}$ and $\mathbf{Y}$, are twofold:

- the expressions for $\mathbf{G_Y}$ and $\mathbf{G_X}$ in terms of $\mathbf{V}$ tend to be simple, often far simpler
  than the original $\mathbf{F}$ functions; and
- there are fewer elements in $\mathbf{V}$ than in $\mathbf{X}$ and $\mathbf{Y}$ (e.g., instead of storing prices and
  quantities separately, we store merely their products, the values of commodity
  or factor flows).

5.4 Levels and Linearized Systems Compared

To illustrate the convenience of the linear approach,[2] we consider a very small
equation system: the CES input demand equations for a producer who makes output
Z from N inputs X_k, $k = 1 - N$, with prices P_k. In the levels the equations are:

---

[2] For a comparison of the levels and linearized approaches to solving AGE models see Hertel
et al. (1992).

$$X_k = Z\delta_k^{1/(\rho+1)} \left[\frac{P_k}{P_{ave}}\right]^{-1/(\rho+1)}, \quad k = 1, \ldots, N \tag{5.8}$$

where

$$P_{ave} = \left(\sum_{i=1}^{N} \delta_i^{1/(\rho+1)} P_i^{\rho/(\rho+1)}\right)^{(\rho+1)/\rho} \tag{5.9}$$

The δ_k and ρ are behavioral parameters. To solve the model in the levels, the values of the δ_k are normally found from historical flows data, $V_k = P_k X_k$, presumed consistent with the equation system and with some externally given value for. This process is called calibration. To fix the X_k, it is usual to assign arbitrary values to the P_k, say 1. This merely sets convenient units for the X_k (base-period-CNY-worth). ρ is normally given by econometric estimates of the elasticity of substitution, $\sigma = 1/(\rho + 1)$. With the P_k, X_k, Z and ρ known, the δ_k can be deduced.

In the solution phase of the levels model, δ_k and ρ are fixed at their calibrated values. The solution algorithm attempts to find P_k, X_k and Z consistent with the levels equations and with other exogenous restrictions. Typically this will involve repeated evaluation of both (1) and (2)—corresponding to F(Y, X)—and of derivatives which come from these equations—corresponding to F_Y and F_X.

The percentage-change approach is far simpler. Corresponding to (5.8) and (5.9), the linearized equations are:

$$X_k = Z - \sigma(P_k - P_{ave}) \quad k = 1, \ldots, N \tag{5.10}$$

and

$$P_{ave} = \sum_{i=1}^{N} S_i P_i \tag{5.11}$$

where the S_i are cost shares, e.g.,

$$S_i = V_i / \sum_{k=1}^{N} V_k \tag{5.12}$$

Since percentage changes have no units, the calibration phase, which amounts to an arbitrary choice of units, is not required. For the same reason the k parameters do not appear. However, the flows data V_k again form the starting point. After each change they are updated by:

$$V_{k,new} = V_{k,old} + V_{k,old}(X_k + P_k)/100 \qquad (5.13)$$

GEMPACK is designed to make the linear solution process as easy as possible. The user specifies the linear equations (5.10), (5.11) and (5.12) and the update formulae (5.13) in the TABLO language—which resembles algebraic notation. Then GEMPACK repeatedly:

- evaluates G_Y and G_X at given values of V;
- solves the linear system to find y, taking advantage of the sparsity of G_Y and G_X; and
- updates the data coefficients V.

The housekeeping details of multistep and extrapolated solutions are hidden from the user. Apart from its simplicity, the linearized approach has two further advantages.

- It allows free choice of which variables are to be exogenous or endogenous. Many levels algorithms do not allow this flexibility.
- To reduce AGE models to manageable size, it is often necessary to use model equations to substitute out matrix variables of large dimensions. In a linear system, we can always make any variable the subject of any equation in which it appears. Hence, substitution is a simple mechanical process. In fact, because GEMPACK performs this routine algebra for the user, the model can be specified in terms of its original behavioral equations, rather than in a reduced form. This reduces the potential for error and makes model equations easier to check.

5.5 The Initial Solution

Our discussion of the solution procedure has so far assumed that we possess an initial solution of the model—$\{Y^0, X^0\}$ or the equivalent V^0—and that results show percentage deviations from this initial state.

In practice, the ORANI database does not show the expected state (B) of the economy at a future date. Instead the most recently available historical data (A) are used. At best, these refer to the present-day economy. Note that, for the temporal static model, A provides a solution for period of T years' time. In the static model, setting all exogenous variables at their base-period levels would leave all the endogenous variables at their base-period levels. Nevertheless, A may not be an empirically plausible control state for the economy at period T and the question therefore arises: Are estimation of the B-to-C (C stands for the state with a shock) percentage changes much affected by starting from A rather than B? For example, would the percentage effects of a tariff cut inflicted in 1994 differ much from those caused by a 2005 cut? Probably not. First, balanced growth, i.e., a proportional enlargement of the model database, just scales equation coefficients equally; it does not affect ORANI results. Second, compositional changes, which do alter percentage-change effects, happen quite slowly. So for short- and medium-run simulations A is a reasonable proxy for B (Dixon et al. 1986).

In this section we provide a formal description of the linear form of the model. Our description is organized around the TABLO file which implements the model in GEMPACK. We present the complete text of the TABLO Input file divided into a sequence of excerpts and supplemented by tables, figures and explanatory text.

The TABLO language in which the file is written is essentially conventional algebra, with names for variables and coefficients chosen to be suggestive of their economic interpretations. Some practice is required for readers to become familiar with the TABLO notation but it is no more complex than alternative means of setting out the model—the notation employed in DPSV (1982), for example. Acquiring the familiarity allows ready access to the GEMPACK programs used to conduct simulations with the model and to convert the results to human-readable form. Both the input and the output of these programs employ the TABLO notation. Moreover, familiarity with the TABLO format is essential for users who may wish to make modifications to the model's structure.

Another compelling reason for using the TABLO Input file to document the model is that it ensures that our description is complete and accurate: complete because the only other data needed by the GEMPACK solution process is numerical (the model's database and the exogenous inputs to particular simulations); and accurate because GEMPACK is nothing more than an equation solving system, incorporating no economic assumptions of its own.

We continue this section with a short introduction to the TABLO language—other details may be picked up later, as they are encountered. Then we describe the input–output database which underlies the model. This structures our subsequent presentation.

5.6 The Key Function of Water and Land Accounts

The model allows land to move between the cultivated land, pasture, and forestry categories, or for unused land to convert to one of these three. A transition matrix show land use changes in the first year of our simulation. The transition matrices (Table 5.2) could be expressed in share form, showing Markov probabilities that a particular hectare used today for, say, Pasture, would next year be used for crops. In the model, these probabilities or proportions are modeled as a function of land rents, via:

$$S_{pq} = \mu_p L_{pq} p l_q^\alpha M_q \tag{5.14}$$

$$A_{ji} = \lambda_i k_{ji} R_{ji}^{0.5} \tag{5.15}$$

Table 5.2 A sample of transition matrices for land use change

	Crop	Pasture	Plant forest	Unused	Total
Crop	X1	X2	X3	X4	SUM(X1:X4)
Pasture	Y1	Y2	Y3	Y4	SUM(Y1:Y4)
PlantForest	Z1	Z2	Z3	Z4	SUM(Z1:Z4)
Unused	M1	M2	M3	M4	SUM(M1:M4)
Total	SUM(X1:M1)	SUM(X2:M2)	SUM(X3:M3)	SUM(X4:M4)	

where,

S_{pq}	share of land type p that becomes type q
μ_p	a slack variable, adjusting to ensure that $\sum_q S_{pq} = 1$
L_{pq}	a constant of calibration = initial value of S_{pq}
Pl_q	average unit rent earned by land type q
α	a sensitivity parameter, with value set to 0.35
M_q	a shift variable, initial value 1
A_{ji}	the area of land type i in region r used for industry j
R_{ji}	the unit land type i rent earned by industry j
K_{ji}	a constant of calibration while the slack variable λ_r adjusts
$\sum_j A_{ji} = A_i$	exogenous area of land type i, including cultivated land, forestry area, industrial land and pasture area.

According the GTAP-W notation and using Eqs. (5.13) and (5.14), the nested tree structure is represented as follows in Fig. 5.7 (we only focus on the value-added nest—where all changes made in GTAP-W take place)

Lower level, first nest: Producers combine irrigable land and irrigation water according to a CES function with elasticity of substitution $ELLW_{j,r}$ (σLW). At this stage, only biased technical change is specified.

Demand for irrigable land (Lnd) and water (Wtr):

$$qfe_{i,j,r} = -afe_{i,j,r} + qlw_{j,r} - ELLW_{j,r} \times [pfe_{i,j,r} - afe_{i,j,r} - plw_{j,r,i}] = Lnd, Wtr \tag{5.16}$$

Unit cost of the irrigable land-water composite:

$$plw_{j,r} = \sum_{k \in ENDWLW} SLW_{k,j,r} \times (pfe_{i,j,r} - afe_{k,j,r}) \tag{5.17}$$

Lower level, second nest: Producers combine capital and the energy composite according to a CES function with elasticity of substitution $ELKE_{j,r}$ (σKE). At this stage, only biased technical change is specified.

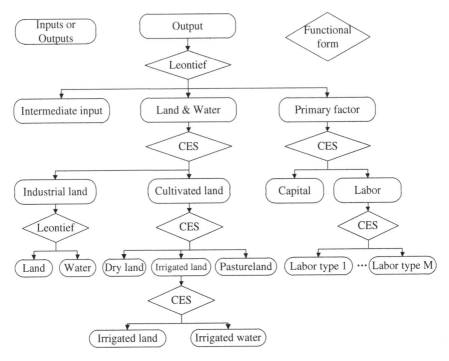

Fig. 5.7 Production function for embed the water and land resource factors

Demand for capital (*Capital*) and the energy composite:

$$afe_{i,j,r} = -afe_{i,j,r} + qke_{j,r} - ELKE_{j,r} \times [pfe_{i,j,r} - afe_{i,j,r} - pke_{j,r,i}] = Capital$$
$$(5.18)$$

$$qen_{j,r} = qke_{j,r} - ELKE_{j,r} \times [pen_{j,r} - pke_{j,r}]$$

Unit cost of the capital-energy composite:

$$pke_{j,r} = \sum_{k \in ENDWC} SKE_{k,j,r} * (pfe_{k,j,r} - afe_{k,j,r}) + \sum_{k \in EGY} SKE_{k,j,r} * (pf_{k,j,r} - af_{k,j,r})$$
$$(5.19)$$

land, natural resources, labor and the "capital-energy" composite according to a CES function with elasticity of substitution $ESUBVA_j$ (σVAE). At this stage, only biased technical change is specified.

Demand for rainfed land (*RfLand*), pasture land (*PsLand*), natural resources (*NatRes*) and labour (*Lab*):

$$qfe_{i,j,r} = -afe_{i,j,r} + qvaen_{j,r} - ESUBVA_j * [pfe_{i,j,r} - afe_{i,j,r} - pvaen_{j,r}] \quad (5.20)$$

Demand for the irrigable land-water composite:

$$i = RfLand, PsLand, NatRes, Lab \tag{5.21}$$

$$qlw_{j,r} = qaven_{j,r} - ESUBVA_j * (plw_{j,r} - qaven_{j,r}) \tag{5.22}$$

Demand for the capital-energy composite:

$$qke_{j,r} = qaven_{j,r} - ESUBVA_j * (pke_{j,r} - paven_{j,r}) \tag{5.23}$$

Unit cost of the value-added composite (including energy inputs):

$$
\begin{aligned}
paven_{j,r} = &\sum_{k \in ENDW} SVAEN_{k,j,r} * (pfe_{k,j,r} - afe_{k,j,r}) \\
&+ \sum_{k \in EGY} SVAEN_{k,j,r} * (pf_{k,j,r} - af_{k,j,r})
\end{aligned}
\tag{5.24}
$$

Upper level: Producers combine the value-added composite with all other inputs according to a CES function with elasticity of substitution $ESUBT_j$ (σ). At this stage, factor biased and neutral technical changes are specified.

Demand for the value-added composite (including energy inputs):

$$qaven_{j,r} = -ava_{j,r} + qo_{j,r} - ao_{j,r} - ESUBT_j * [paven_{j,r} - ava_{j,r} - ps_{j,r} - ao_{j,r}] \tag{5.25}$$

Demand for all other inputs (excluding energy inputs but including energy feedstock):

$$
\begin{aligned}
qf_{i,j,r} = &\,D\_NEGY_{i,j,r} \times D\_VFA_{i,j,r} \times [-af_{i,j,r} + qo_{j,r} - ao_{j,r} - ESUBT_j \times (pf_{i,j,r} - af_{i,j,r} - ps_{j,r})] \\
&+ D\_ELY_{i,j,r} \times D\_VFA_{i,j,r} \times [-af_{i,j,r} + qen_{j,r} - ELELY_{j,r} \times (pf_{i,j,r} - af_{i,j,r} - pnel_{j,r})] \\
&+ D\_COAL_{i,j,r} \times D\_VFA_{i,j,r} \times [-af_{i,j,r} + qenl_{j,r} - ELCO_{j,r} \times (pf_{i,j,r} - af_{i,j,r} - pnel_{j,r})] \\
&+ D\_OFF_{i,j,r} \times D\_VFA_{i,j,r} \times [-af_{i,j,r} + qncoal_{j,r} - ELFU_{j,r} \times (pf_{i,j,r} - af_{i,j,r} - pncoal_{j,r})]
\end{aligned}
\tag{5.26}
$$

Unit cost of the output:

$$
\begin{aligned}
ps_{j,r} + ao_{j,r} = &\sum_{i \in ENDW} STC_{i,j,r} * [pfe_{i,j,r} - afe_{i,j,r} - ava_{j,r}] \\
&+ \sum_{i \in TRAD} STC_{i,j,r} * [pf_{k,j,r} - af_{k,j,r}] + profitslack_{j,r}
\end{aligned}
\tag{5.27}
$$

where,

$qfe_{i,j,r}$	demand for endowment i for use in industry j in region r
$qlw_{j,r}$	composite "irrigable land + water" in industry j of region r

$qke_{j,r}$	composite "capital + energy" in industry j of region r
$qen_{j,r}$	composite energy (electricity + non-electricity) in industry j of region r
$qvaen_{j,r}$	value-added in industry j of region r
$qo_{i,r}$	industry output of commodity i in region r
$qf_{i,j,r}$	demand for commodity i for use by j in region r
$qnel_{j,r}$	composite non-electric good in industry j of region r
$qncoal_{j,r}$	composite non-coal energy good in industry j of region r
$pfe_{i,j,r}$	firms' price for endowment commodity i in industry j of region r
$plw_{j,r}$	firms' price of "irrigable land + water" composite in industry j of region r
$pke_{j,r}$	firms' price of "capital + energy" composite in industry j of region r
$pen_{j,r}$	price of energy (elec. + non-elec.) composite in industry j of region r
$pf_{i,j,r}$	firms' price for commodity i for use by industry j in region r
$pvaen_{j,r}$	firms' price of value-added in industry j of region r
$ps_{i,r}$	supply price of commodity i in region r
$pnel_{j,r}$	price of non-electric composite in industry j of region r
$pncoal_{j,r}$	price of non-coal composite in industry j of region r
$afe_{i,j,r}$	primary factor i augmenting technical change by industry j of region r.

5.7 The TABLO Language

The TABLO model description defines the percentage-change equations of the model.

For example, the CES demand Eqs. (5.8) and (5.9), would appear as:

```
Equation E_x # input demands #
  (all, f, FAC)   x(f) = z - SIGMA*[p(f) - p_f];
Equation E_p_f # input cost index #
  V_F*p_f = sum{f,FAC, V(f)*p(f)};
```

The first word, 'Equation', is a keyword which defines the statement type. Then follows the identifier for the equation, which must be unique. The descriptive text between '#' symbols is optional—it appears in certain report files. The expression '(all, f, FAC)' signifies that the equation is a matrix equation, containing one scalar equation for each element of the set FAC.[3]

Within the equation, the convention is followed of using lower-case letters for the percentage-change variables (x, z, p and p_f), and upper case for the coefficients (SIGMA, V and V_F). Since GEMPACK ignores case, this practice assists only the human reader. An implication is that we cannot use the same sequence of characters, distinguished only by case, to define a variable and a coefficient. The '(f)' suffix indicates that variables and coefficients are vectors, with elements corresponding to the set FAC. A semicolon signals the end of the TABLO statement.

To facilitate portability between computing environments, the TABLO character set is quite restricted—only alphanumerics and a few punctuation marks may be available. The use of Greek letters and subscripts is precluded, and the asterisk, '*', must replace the multiplication symbol ' $\times$ '.

Sets, coefficients and variables must be explicitly declared, via statements such as: As the last two statements in the 'Coefficient' block and the last three in the

```
Set FAC # inputs # (capital, labour, energy);
Coefficient
(all,f,FAC) V(f) # cost of inputs #;
         V_F      # total cost #;
         SIGMA      # substitution elasticity #;
Variable
(all,f,FAC) p(f) # price of inputs   #;
(all,f,FAC) x(f) # demand for inputs   #;
         z      # output #;
         p_f      # input cost index #;
```

---

[3] For equation E_x we could have written: (all, j, FAC) $x(j) = z - SIGMA * [p(j) - p\_f]$, without affecting simulation results. Our convention that the index (f), be the same as the initial letter of the set it ranges over, aids comprehension but is not enforced by GEMPACK. By contrast, GAMS (a competing software package) enforces consistent usage of set indices by rigidly connecting indices with the corresponding sets.

'Variable' block illustrate, initial keywords (such as 'Coefficient' and 'Variable') may be omitted if the previous statement was of the same type.

Coefficients must be assigned values, either by reading from file:

Read V from file FLOWDATA;

Read SIGMA from file PARAMS;

or in terms of other coefficients, using formulae:

Formula V_F = sum{f, FAC, V(f)};   ! used in cost index equation !

The right hand side of the last statement employs the TABLO summation notation, equivalent to the notation used in standard algebra. It defines the sum over an index f running over the set FAC of the input-cost coefficients, V(f). The statement also contains a comment, i.e., the text between exclamation marks (!). TABLO ignores comments.

Some of the coefficients will be updated during multistep computations. This requires the inclusion of statements such as:

Update (all, f, FAC) V(f) = x(f)*p(f);

---

which is the default update statement, causing V(f) to be increased after each step by [x(f) + p(f)] %, where x(f) and p(f) are the percentage changes computed at the previous step.

The sample statements listed above introduce most of the types of statement required for the model. But since all sets, variables and coefficients must be defined before they are used, and since coefficients must be assigned values before appearing in equations, it is necessary for the order of the TABLO statements to be almost the reverse of the order in which they appear above. The MODEL TABLO Input file is ordered as follows:

- definition of sets;
- declarations of variables;
- declarations of often-used coefficients which are read from files, with associated Read and Update statements;
- declarations of other often-used coefficients which are computed from the data, using associated Formulae;
- groups of topically-related equations, with some of the groups including statements defining coefficients which are used only within that group.

References

Deng, X. (2011). *Environmental computable general equilibrium model and its application.* Beijing: Science Press. (in Chinese).

Dixon, P. B., Parmenter, B. R., & Rimmer, R. J. (1986). ORANI projections of the short-run effects of a 50 per cent across-the-board cut in protection using alternative data bases. In J. Whalley & T. N. Srinivasan (Eds.), *General equilibrium trade policy modelling* (pp. 33–60). Cambridge, MA: MIT Press.

Hertel, T. W., Horridge, J. M., Pearson, K. R., (1992). Mending the family tree a reconciliation of the linearization and levels schools of AGE modelling. *Economic Modelling, 9*(4), 385–407.

Johansen, L. A. (1960). Multisectoral study of economic growth. Amesterdam: North-Holland Publishing Company.

Qureshi, M. E., Proctor, W., Young, M. D., & Wittwer, G. (2012). The economic impact of increased water demand in Australia: A computable general equilibrium analysis. *Economic Papers: A journal of applied economics and policy, 31*(1), 87–102.